3-14-77

W. Blasius

Problems of Life Research

Physiological Analyses and
Phenomenological Interpretations

With 45 Figures

Springer-Verlag
Berlin Heidelberg New York 1976

Original title of the German edition:
"Probleme der Lebensforschung".
Copyright © 1973 by Rombach + Co GmbH, Freiburg

Wilhelm Johann Heinrich Blasius
Zentrum für Physiologie, Bereich Humanmedizin
der Justus Liebig Universität, Abteilung für
Angewandte Physiologie, Aulweg 129, 6300 Gießen

ISBN 3-540-07731-6 Springer-Verlag Berlin · Heidelberg · New York
ISBN 0-387-07731-6 Springer-Verlag New York · Heidelberg · Berlin

Library of Congress Cataloging in Publication Data. Blasius, Wilhelm,
1913-. Problems of life research. Translation of Probleme der Lebensforschung. Bibliography: p. Includes index. 1. Biology-Philosophy.
I. Title. [DNLM: 1. Physiology. QT104 B644p] QH331.B54413
574'.01 76-12466

This work is subject to copyright. All rights are reserved, whether the
whole or part of the material is concerned, specifically those of
translation, reprinting, re-use of illustrations, broadcasting, reproduction
by photocopying machine or similar means, and storage in data banks.
Under § 54 of the German Copyright Law, where copies are made for
other than private use, a fee is payable to the publisher, the amount
of the fee to be determined by agreement with the publisher.

© by Springer-Verlag Berlin · Heidelberg 1976
Printed in Germany.

The use of registered names, trademarks, etc. in this publication does
not imply, even in the absence of a specific statement, that such names
are exempt from the relevant protective laws and regulations and
therefore free for general use.
Typesetting, printing and binding: fotokop wilhelm weihert KG, Darmstadt.

Foreword

Professor *Wilhelm Blasius*, physiologist at Giessen in West Germany, has written a book "Probleme der Lebensforschung" (Verlag Rombach, Freiburg 1973) which – I understand – is to be published in an English version.
To me it has been of interest as an orientation in a world of traditional German thinking, best known from *Goethe's* natural philosophy of perceptible "Urbilder", which perhaps in English could be rendered descriptively by calling it an inner vision of further irreducible totalities. It is in contemporary language a kind of intuitive 'holistic' insight which represents understanding different from that of natural science. The latter is devoted to the study of causal chains, is largely experimental and in its aim ultimately 'reductionist' – to use another modern term. *Goethe's* approach is reincarnated in the "Wesenslehre" of the late *Ludwig Klages* (1872–1956) and in the thinking of *Carl Gustav Carus*. To *Klages* the perceived image alone is the meaning of everything in this world and in this sense he is the advocate of a psychological phenomenology that is an end in itself. Originally trained as a chemist *Klages* soon turned to philosophy and developed a system of concepts supposed to give a deeper insight into the essence of life than did the endless causal chains of natural science. *Blasius* makes much use of the concepts of this philosopher, in particular his ideas on 'polarities'. At its best the idea of 'polarity' is identical with the 'complementarity' of *Niels Bohr* which refers to approaches that supplement one another like the corpuscular and the

wave theory of light but 'polarity' has numerous other applications throughout *Blasius'* work. These I have been unable to follow or to see a need for. Although *Blasius* himself has contributed serious experimental work, e. g. to cardiology, the other side of his Janus face, turned towards *Klages,* makes him quote a number of experiments to illustrate that the endless multiplicity of causes has lead to an inherent and disappointing aimlesness of modern science in its mass production of quantified detail. The now so popular molecular biology held by some to lead to final explanations is critisized by *Blasius* for the same reason; the deeper the penetration into the molecular events, the less the degree of predictability, as such a consequence of causal understandig. In several chapters *Blasius* devotes himself to popular biology taking up German work, partly his own, which is less known to English-speaking scientists. His popular writing is easily followed as he presents a number of interesting observations on biological rhythms, open and closed systems, effects of exercise, correlations between *Kretschmer's* constitutional types on the one hand, relative sensitivity to form or colour on the other. One amusing chapter discusses sensible and nonsensical consequences of thinking in terms of numerical correlations ("Zifferdenken"). The examples are curves showing the expansion of world population, of number of journals devoted to the science of physiology, the increase in speed of vehicles of transport, the rising number of deaths in traffic accidents, of mathematical expressions in science etc.

Blasius' two approaches to biological insight, the scientific and that of natural philosophy, do not blend but preserve absolute independence. The latter involves a considerable element of conviction and in this sense his book has a confessional aspect. Others may write popular biology with equal facility but few of them would have dared to rise in defense of an essentially humanistic attitude toward biological understanding.

RAGNAR GRANIT

Foreword to the Second Edition and English Translation

Both inside and outside the German-speaking world "Probleme der Lebensforschung" received unexpected acclaim from physiologists, morphologists, biologists, psychologists, physicists and philosophers. The English translation was undertaken in order to make this book accessible to a still larger circle of readers. Thanks are due to Springer-Verlag, Berlin–Heidelberg–New York for making publication of the book in English possible. Without an appropriate translation of concepts such as "Polarität", "Erlebnis", "Schauen", "Gestalt", "Seele", "Bild", "Erscheinung", "Wesen" und "Geist", "Auffassungsakt", "Ding", "Dualität" etc., the philosphical part of the book would have remained incomprehensible and inaccessible. In view of this I should like to explain in this foreword at least the most important of these concepts.

"Polarität" is a fundamental concept in natural philosophy. It has stirred Western thought since *Heraclitus* and also forms the basis of the natural philosophical works of *Goethe, Carus* and *Klages*. It concerns the coherence of the physical and psychic poles of a living creature or generally of two interconnected domains of life, such as body and soul, the left and right side of the body, male and female being and also movement and rest, light and shade, pleasure and pain etc.

"Schau" or "schauen" is the intuitive holistic apprehending of a "Gestalt", a phenomenon ("Erscheinung"), a living image ("Bild"), i.e. of an essence ("Wesen"). This figurative ("gestalthafte"), holistic apprehension of essence must not be confused with sensation ("Empfindung"), which is merely concerned with the physical processes of sensory excitation.

The *antithesis between the soul ("Seele") and spirit ("Geist") of man* is of great importance. While the human soul can be understood above all as ability to apprehend and form ("gestalten"), the spirit refers to the cognitive powers which enable conscious acts of cognition ("Auffassungsakt") and to the capacity of will which moves the bearer of spirit to acts of will ("Willensakt"). There is a polar coherence between soul and body, but an antithesis between spirit and soul; the latter is distinguished from polarity as *duality*.

The English words chosen for these concepts do not always correspond exactly to the content and meaning of the German expressions. However, they were regularly adhered to in order not to confuse the reader and not to impair his understanding for the whole of the book. In cases of doubt, the original works quoted give further information.

It will certainly be valuable for the reader to hear some assessments of the value and limitations of the causal-analytical principle in the natural sciences. Among these, the judgements of the theoretical physicists should be of particular significance: they are impartial compared to the physiologists and biologists; and physics constitutes the top of the whole development in the natural sciences. For physicists criticism of their own results is one of the highest virtues.

On the basis of quantum theoretical considerations the theoretical physicist *Heisenberg* arrived at the 'uncertainty principle' named after him. This principle describes the statistical character of all processes at the atomic level. With this statistical interpretation of processes in atoms, a purely causal analysis had to be given up once and for all in this sphere. Of course, the causality of 'classical' physics in the areas in which it is valid was not infringed by the *Heisenberg* Principle.

Something quite similar is now occuring in biology. The purely causal analysis which has been in the ascendancy in all fields of biology since the last century also cannot be applied to all domains of life. The analytical procedure must hence be supplemented by a figurative, intuitive apprehension of the living phenomena.

The theoretical physicist *Heitler* (Zurich) has already done siginificant work in this field by advocating such a comprehensive view of everything living. In his book 'Man and Insight in the Natural Sciences' (Der Mensch und die naturwissenschaftliche Erkenntnis, 4th edition, Brunswick 1966) he says: 'The prevailing natural science distances itself more and more from life and from man. It is consciously nonanthropomorphic... How much is lost through restriction to the quantitative and to the causal sequence? Practically everything that we really *experience:* sensory qualities, color, sounds, smell etc., all which have Gestalt and form, above all the Gestalt of living organisms'. These views (expressed by a physicist!) are fully shared by the author of the present book, which is largely concerned with just these problems. It was a great satisfaction for me that *Heitler* expressed in a letter his full agreement with my views and efforts.

I place just as much value on the opinions of biologists. Above all the excellent study by *Reinle* (Basel) 'The World of Experience and the World of Science' (Erlebniswelt und Wissenschaftswelt, Hestia, Bonn 1975) should be mentioned here. Quite similar to the ideas presented in my book, it is devoted to the theme of the twofold relation of man to reality. I wish to mention the agreement expressed by the zoologist *Ankel* (Giessen), the physiologist *Lullies* (Kiel), the psychiatrist *Shichi-ro Chidani* (Tokyo), the internist *Kötschau* (Rosenheim), the morphologist *Kaiser* (Washington), the philosopher *Funke* (Mainz), the psychologists *Wellek* (Mainz) and H. A. *Müller* (Würzburg) and many others, and take this opportunity to express my sincere thanks.

Particular thanks are due to the esteemed neurophysiologist *Ragnar Granit* (Stockholm) who has written the preface for the English edition of my book. These words of the world-renowned master in physiology will be a lasting stimulus to continue along the lines followed and further develop the book.

All the gratifying testimony of the contemporaries falls in with a chorus of voices which ring out a hymn to nature and her ingenious activity. The purest and clearest of these voices is that of the natural researcher and poet *Johann Wolfgang von Goethe*. In his conversation with *Eckermann* on 13th February 1829 he spoke as follows:

"But nature knows no fun; she is always true, serious, stern. She is always right; faults and errors are always man's. She scorns the inadequate; she surrenders and reveals her secrets to the competent, genuine and pure.

The intellect does not reach up to her; man must be capable of elevating himself to the heights of reason in order to sense the godhead that manifests herself in those archetypal phenomena which derive from her and behind which she stands.

But the godhead is active in the living, not in the dead; her essence lies in evolving and metamorphosing, not in what once was and is now rigid. Therefore, in her tendency to the divine, reason concerns herself only with the evolving, the living; the intellect with that which was, the static, in order to make use of it".

May these words accompany the book into the world as a seal of truth and gain it many new friends.

Spring of 1976   W. Blasius

Preface to the First Edition

> Living means subsisting on an enveloping world which we feel or suspect to be an irreducible whole.
> Ortega y Gasset,
> "What is Philosophy?"

Research in the life sciences has not always been orientated so one-sidedly as today towards determing quantitative data and towards causal analysis of isolated life processes. In ancient times there were observers of nature (there were also such observers in the medieval and modern periods) who attempted to describe life in all its appearences and forms (Gestalten), movements, rhythms and transformations, polarities, analogies and manifold connections. Man was also integrated in the rhythm of the whole living world.

Withdrawal of man from these (vivid connections), his intellectual isolation, led (or indeed forced) him into an antithetical position to nature as a whole. This was the real reason why only experimental data, numerical values and facts retained validity in biological research. This corresponded to his rational inclination and gave man the ability to transform nature. At first the transformation was gentle so as not disturb or totally destroy nature. However, this tendency to transform nature eventually became overpowering and man began to exploit nature and non-human life to an excessive extent and finally to exhaust it.

All insights in the field of biology hence caution serious reflection and warn against a purely utilitarian way of thinking in investigation and evaluation of life. The majority of physiologists (but also biologists, natural scientists and psychologists) nevertheless continue to work with the causal-analytical concept.

In this situation it appears urgently necessary to undertake a clarification of the various terms and the fundamental antithesis of phenomenological and analytical ways of thinking. Above all the limitations of the two modes of thinking must be described; it is not a matter of disparaging one of them and its methods but rather of drawing anew the visual manifestations of life and depict their meaning for the vital connections of *all* living entities and *all* domains of living nature.

I have long viewed such a description as an important, indeed a vital, task. It was satisfying to have found a great deal of sympathy for this idea. Particular thanks are due to Mr. *Kaltenbrunner*, Editor with Rombach Publishers now Herder Publishers Freiburg, who offered me an opportunity to summarize the results of many years of research which I had in part already published in a textbook, in various specialist journals and in other publications. Diagrams were necessary to illustrate many situations and developments. I hence also thank the Publisher for this additional expenditure. I hope that all the physiologists, biologists, medical researchers and psychologists who are concerned with these problems will be stimulated to further studies in the direction indicated. Their combined efforts may help to maintain life on earth, to promote it and, where life is expiring under human pressure, to renew it again. Only such aspirations can constitute the meaning and objective of *Research of Life* in its truest sense, i.e. *Research in the Service of Life*.

Spring of 1973   W. Blasius

Contents

I. Epistemological and Methodological Foundations of Research into Life 1

1. Introduction 1
1.1. Historical Review of Modification in the Concept "Physiology" 2
2. Epistemological Foundations of the Science of Life 3
2.1. The Natural Philosophical or Holistic Theory of Life 5
2.2. The Natural Scientific or Reductionist (Teilinhaltliche) Theory of Life 10
2.3. Limitations of the Two Lines of Thought 13
3. The Scope of the Natural Scientific Theory of Life 21
3.1. Methodological Foundations of Physiology as Natural Science 22
3.2. Natural Laws and Rules 23
3.3. Limits of a Metrical and Mathematical Treatment of the Phenomena of Life 24
3.4. Mathematical Methods 26
3.5. Parameters 27
3.6. Validity of Laws 28
4. Special Attributes of Organisms 28
4.1. Cellular Organization 29
4.2. Chemical Constitution of Organisms 29
4.3. Transformation of Energy 30
4.4. Relations with the Environment 31
4.5. Evolution of Life 32
4.6. The Course of Life 34

- 4.7. Excitability 34
- 4.8. Capacity for Regulation 35
- 4.9. Animation 35
- 5. The Flow of Energy as the Most Important Principle of Scientific Biology 36
- 5.1. The Energetics of Closed Physical Systems 36
- 5.2. The Organism as an Open System 39
- 5.3. General Properties of Open Systems 40
- 5.4. Kinetics of Open Systems 43
- 5.5. The Thermodynamics of Open Systems 44
- 5.6. Provenance of the Energy for Living Processes 46
- 6. Summary 48

II. Rhythm and Polarity – Physiological Analysis and Phenomenological Interpretation 49

- 1. Causal Analysis of Periodic Processes 50
- 1.1. Phases of Rest and Activity in Animals 50
- 1.2. Phases of Activity and Rest in Man 56
- 2. A Critical Analysis of a Causal-Analytic Interpretation of Rhythmic Phenomena 59
- 3. Phenomenological Description of Rhythm and Polarity 64
- 4. Summary 68

III. Bodily Movement and Exercise – Physiological Analysis and Phenomenological Interpretation 69

- 1. Phenomenological Interpretation of Bodily Movement 69
- 2. Physiological Analysis of Bodily Movements 70
- 2.1. The External Movement of the Organism in the Interplay of Muscles and Nervous System 71
- 2.1.1. The Basic Properties of Muscles 71

- 2.2. The Role of the Muscle Spindles in the Interplay of Muscles and Nervous System 75
- 2.3. Physiological and Morphological Changes in the Organism during Physical Exercise 79
- 2.4. Muscular Adaptation 81
- 2.5. Adaptation of Heart Muscle 83
- 2.6. Adaptation of the Blood 85
- 2.7. Functional Adaptation 86
- 2.7.1. Circulation 86
- 2.7.2. Respiration 86
- 2.7.3. Utilization 87
- 2.7.4. Autonomic Nervous System 88
- 3. Exercise as a Therapeutic Measure 89
- 4. Summary 90

IV. Human Language – Physiological Analysis and Phenomenological Interpretation 92

- 1. Physiological Analysis of the Sound of Speech and of Hearing 95
- 1.1. The Activity Cycle of Hearing and Speech 95
- 1.2. The Theory of Centers and Plasticity for the Explanation of Language 98
- 1.3. Electroencephalographic Results of Excitation of the Cerebral Cortex through Optical and Acoustic Stimuli 104
- 1.4. Electroencephalographic Observations during Mental Activity 106
- 2. A Phenomenological View of Language 107
- 2.1. Soul and Spirit of Language 107
- 2.2. The Contrast between Linguistic Meanings and Linguistic Concepts 112
- 2.3. The Difference between Meaning and Conception Words 116
- 3. Summary 120
- 4. Postscript 122

V. Stereoscopic Vision and Color Discrimination: Their Typological Polarity and Relations to Pictorial Creativeness   123

1. Physiological Methods   124
1.1. Testing Stereoscopic Vision   124
1.2. Testing the Capacity to Discriminate Color   128

2. Psychological Methods   129
2.1. Investigation of Color Selection   129
2.2. The Creativity Test   130
2.3. The Self-Diagnosis Test   130
2.4. The Experimental Material   131

3. Performance Differences in Stereoscopic Vision   131
4. Performance Differences in Color Discrimination   132
5. Typological and Sex Differences in the Kind of Creativity   132
6. Stereoscopic Vision and Color Discrimination in Relation to Kind of Creativity   134
7. The Complementary Behavior of Stereoscopic Visual Capacity and Color Discrimination Ability   136
8. Color Discrimination, Stereoscopic Vision and the Kind of Creativity in Relation to Color Selection   138
9. On the Polarity of Form and Color Perception   140
10. Color Selection as a Polar Phenomenon   144
11. Summary   146

VI. Quantitative Thinking in the Life Research   147

VII. The Essence of Health and Its Maintenance   169

VIII. Bibliography   186

IX. Author Index   193

X. Subject Index   195

I. Epistemological and Methodological Foundations of Research into Life

> The whole cognitive apparatus is not
> directed towards understanding but
> towards coping with the universe.
> Nietzsche

1. Introduction

Content and purpose of present-day research in the life sciences can be understood in historical terms and illustrated by analyzing its epistemological basis.

Speculation on the general nature of vital processes and the special feats of human and animal organs is as old as science itself. It preoccupied philosophers as much as scientists and physicians from time immemorial. The answers to this enquiry are of great consequence for every philosophy of man as well as for the practical reasoning and work of the physician.

The scope of physiology as a life science has been differently construed in various epochs of cultural and intellectual development; it was accordingly studied by different means and methods at various periods. The current view of physiology - a theory of life orientated to natural science - is only one among many. The reasoning, aims and methods of physiology have changed with advances in knowledge of cosmology, anthropology and philosophy, so it can only be truly understood in relation to the changes in the respective intellectual periods.

1.1. Historical Review of Modifications in the Concept "Physiology"

The diversity of views on the tasks of physiology since *Aristotle* first coined the scientific term clearly shows the lability of its meaning. For *Aristotle*, "physiology" is still natural science in the widest sense, i.e. not restricted to the organic world. The Greek word φωσις means "nature as an ordered whole". For *Hippocrates* and his disciples, the word physiology essentially serves to define the position of man within the universe as a whole. The Hippocratic School used φωσις to designate the "healing force" of nature; this use of the word is more specific than *Aristotle*'s and is also restricted to man. Under *Galen*, physiology became a "theory of the utility of parts and elements" (of the plant, animal or human body). This reductionist ("teilinhaltlich") physiological doctrine dominated scientific thought until late into the Middle Ages. In the Renaissance, *Paracelsus* developed a more holistic ("ganzheitlich") theory of life, though this was not free from all sorts of obscure, mystical and opaque ideas. Under the influence of *Boerhaave* in the era of the Enlightenment and Rationalism, physiology was restricted to the science of the function of organs. Great importance was now attached to the way in which organs work. *Albrecht von Haller* (the distinguished student of *Boerhaave*) described all the processes occurring in the anatomic organ structures in his eight-volume work "Elementa Physiologiae". In the concept of *Goethe*, *Schelling*, *Oken* and *Carus*, physiology was once more incorporated into a general, coherent picture of nature encompassing all its domains and elements; this view resembles that of the Greek natural philosophers *Heraclitus* and *Alkmaion* (as well as that of *Paracelsus*). This concept of physiology was conceived by intuitive philosophers and contemplators of nature. The natural philosophers rejected scientific experimentation because it would falsify the image of nature which people attain. Under the influence of the results of physical and chemical research in the 17th and 18th centuries, physiology became a purely scientific theory of the physico-chemical functioning of the body (especially in the 19th century under the leadership of *Johannes Müller*, *Claude Bernard*, *Robert Mayer*, *Carl Ludwig* and *Hermann von Helmholtz*).

The task of physiology thus changes with the philosophical climate of each epoch (*Rothschuh*).

2. Epistemological Foundations of the Science of Life

There are basically two opposing approaches or doctrines which have given rise to quite different theories of life in the course of history. They can be contrasted as "natural philosophy" [holistic or intuitive-figurative ("intuitiv-bildhaft")] and "natural science" (reductionist = teilinhaltlich or experimental-quantitative). This does not mean that only one of the two approaches is scientifically orientated; the contrasted designations should be understood in terms of the history of science. They reflect the Greek theory of natural philosophy (extended by medieval and modern natural philosophers) on the one hand, and the scientific interpretation of physiology developed in the last century on the other.

It is important to become thoroughly acquainted with these two approaches because conflict between the two arises repeatedly in discussions of the problem of life - its genesis, description, meaning and utilization. It is hence useful to know the conceptual realm one is dealing with.

The object, method, result and aim of the two approaches can be contrasted and elucidated in a simple scheme (Table 1). Typical modern research trends in natural philosophy and natural science can also be contrasted with historical doctrines. The limitations and possible dangers of the two approaches can be illustrated too.

Table 1. Theories of life (after *Blasius*)

	Natural Philosophical Theory of life = Gestalt Theory or Phenomenology	Theory of life based on Natural Science = Causal Theory
Objects	Phenomena (Gestalten), forms images = essences = souls Images (visual-, sound-, tactil-images) Archetypal images (Urbilder) Wholes (Ganzheiten) Qualities Properties = aspects of phenomena	Things as such = conceptualized objects = abstractions of images Facts, forces Causes Units, elements Quantities Numbers
Methods	Perception = perception of essences Conclusion by analogy Figurative (symbolic) Thinking	Abstract consideration and explanation of things Conclusion by causality Logical (mathematical) Thinking
Results	Arrangement according to archetypal images Connections Polarities Knowledge arrived at intuitively Cosmic universality of life	Causal explanation Abstract relations Systems of notions Knowledge arrived at inductively Distinction of dead and living matter
Aims	To interpret the meaning of life Contemplation of the world and the mystery of the cosmos Image of the world	To establish natural laws Domination of the world Collapse of images
Limits	Holistic ("ganzheitlich") world view Meaningful connections and images	Reductionist system Conditionality of natural laws (parameters, constants)

Dangers = transgression of limits	Grouping of "things as such" and facts by analogy and polarity Attempt to interpret unperceptible phenomena (unmanifested life) Blending of "Weltanschauung" with natural scientific results	Analysis of animate phenomena instead of energetic and material things Aimlessness because of the endlessness of the "Chain of causes" Conquest and domination of the world = Decomposition of the image of the world
Examples	Biocentric Thinking Metaphysics Greek natural philosophy (presophists): Heraclitus, Protagoras	Logocentric Thinking Physics Plato, Aristotle
	The Romantic natural philosophy: Goethe, Schelling, Oken, Carus	The Enlightenment and rationalism: Descartes, Newton, Kant
		Physiology as Natural Science: R. Mayer, C. Ludwig, H. v. Helmholtz
	(Vitalism) Phenomenology (Erscheinungswissenschaft)	Mechanism Causal-analytical Science
	Holism (Ganzheitslehre) Nature-cure Characterology Body Language Typology	Separate sciences Scientific medicine Reflex theory Information theory
	Sense research	Purpose research

2.1. The Natural Philosophical or Holistic Theory of Life

The natural philosophical theory of life, also called theory of forms (Gestaltlehre; *Goethe*) or phenomenology (Wesenslehre; *Ludwig Klages*) classifies and describes living phenomena, which are perceived as having form (Gestalt) and soul at the same time. This theory is really a "psychology", a theory of soul and life. The Greek word ψωχη does not only mean "soul"; it also means life (true psychology is therefore also a theory of life). This theory interprets living phenomena as forms or images in which "souls" or "entities" (discrete essences) are expressed. The object of this philosophy includes all real expressions of life and the world, all forms and archetypes, all wholes ("Ganzheiten") as *Goethe* called them. Such images are described according to their perceptible attributes and qualities and integrated into the global image of the universe.

Goethe's morphology can be considered a model exemplifying such a natural philosophical classification of images and forms. Words (spoken by man as an articulate creature) are also images; they have the meaning of entities or souls and represent the response of a person to stimulation by an image. Perception of an image therefore entails simultaneous perception of the essence of an image. Nature is so arranged that man can understand it as a whole by contemplating it; all violent probing and experimentation appears superfluous in view of the readiness of nature to reveal its essential relationships. By analogy, i.e. through the discovery of similarities between images, souls and entities the world is classified; internal and external relationships are discovered.

Certain symbols distinguish images of especially vivid phenomena in which the entities of diverse images coincide; σωμβολον is derived from the Greek verb σωμβαλλειν = coincide and its meaning is roughly "coincident". Such symbols indicate the internal relationships of many phenomena, e.g. the fire, the spring, the river, the leaf, the twig, the tree, the circle, the arrow etc.

The natural polarities seen in the relationships found between living phenomena result from this thinking. Male and female, body and soul, phonemes (as the body of language) and the meaning of language (as its soul), the microcosm and the macrocosm and many other such pairs are polar opposites according to the natural philosophical view. Nature has created everything in polarities which are essential to its continued life.

The view of *Ludwig Klages* (1872-1956) on the connection between body and soul is particularly important in this relation. The connection between soul and body is not the same as that between cause and effect but rather like that between meaning and apparent meaning. This was expressed for the first time by *Carl Gustav Carus* thus: the soul is the meaning of the body phenomenon, and the body is the outward appearance of the soul. The soul neither acts upon the body nor the latter on the former, since neither belong to the world of material things ("Dinge"). If the sequence of cause and effect merely signifies a relationship between parts of a continuum which we have broken up in our thought, then meaning and apparent meaning are themselves a continuum, "the archetype of all continua" as *Klages* so aptly expressed it.

I have made special reference to *Klages* in support of my line of thought because he was himself a natural scientist. He had attended lectures by *Ostwald* (the founder of physical chemistry) in Leipzig and took his doctorate under *Einhorn* in Munich on the basis of experimental work in chemistry. Though his interest in the natural sciences was never extinguished, *Klages* then turned to psychological and philosophical studies. In 1908 he met the Hungarian physicist *Melchior Palágyi* whose theory of cognition became important for the development of *Klages'* philosophy. A lively scientific interchange developed which was ended by the premature death of *Palágyi* in 1924.

I have great confidence in *Klages'* philosophy because of these considerations. He arrived at his comprehensive view of the phenomena of life only after thorough acquaintance with the methods and principles of exact sciences. Thus *Klages* was not educated in the tradition of pure philosophy, but first he became acquainted with the essence, value and limitations of natural science; serious pursuit of natural science contributed substantially to the clarity of his thinking and was a necessary assumption for the breadth of his philosophy. He analyzed the unquestioned accepted assumptions of the natural sciences and considered them against the background of the phenomena of life and living images which were in danger of being ignored in the striving to establish precise conceptual systems. *Klages* may thus be considered one of the most important and far-seeing philosophers of our century.

He made his epistemological contributions to the principles of natural science in 1921 in his fundamental work "On the Essence of Consciousness" (Vom Wesen des Bewußtseins; fourth edition published in 1955, just before his death). This book should be closely studied by every natural scientist because it shows the limitations of a purely scientific treatment of the phenomena of nature and life. *Klages*' comprehensive philosophy accomodates scientific theory but shows very clearly that an analytical-reductionist interpretation of vital processes which does not take account of the world and life as a whole is quite clearly meaningless and without solid foundation. *Klages*' fundamental intellectual achievment and the profound effect of his philosophy consist in his having rescued our thinking from a positivistic drift which had threatened to become excessive throughout the natural sciences and in having redirected our thinking towards the reality of life and the world.

Results of the valuable impulses emanating from him are to be found in many fields. *E. Frauchiger* (1947) emphasized the importance of *Klages*' psychology for biology and medicine; his influence is sure to benefit other natural sciences as well.

In "*Landois-Rosemann*'s " Textbook of Human Physiology (Lehrbuch der Physiologie des Menschen; Vol. 2, 1962), I attempted to contrast the essential features of *Klages*' natural philosophy with the scientific theory of life. *Klages*' theory was deliberately classified in the natural philosophical tradition to which he had always maintained that it belonged: Greek natural philosophy beginning with *Heraclitus* and romantic natural philosophy (including *C.G. Carus* and *Goethe* in particular). The latter (to whom he was greatly indebted) were held in special regard by *Klages*. *Nietzche*'s ideas also profoundly affected his thought. *Klages* revived and completed natural philosophy in the 20th century calling attention to the great Western tradition in this branch of philosophy. The unique character and profound originality of his philosophical achievment should be amply clear.

After this digression on the life and work of *Klages* we return to presenting natural philosophy and shall illustrate with further examples the polar character of life which was one of his preoccupations.

The left and right side of the body, attraction and repulsion, stimulus and excitation, light and shade, the qualities hot and cold, wet and dry, positive electricity and negative electricity, positive magnetism and negative magnetism etc. are striking examples of polarities. Incidentally, the axioms of physics are found among these polarly divergent tendencies; they cannot be further explained by a theory orientated to natural science and have to be accepted as given.

For the natural philosopher, the meaning of life is to perceive real images, to interpret and to mold the meaning of the polar relationships and equivalences connected with images. Developing ideas of *Goethe* and *Nietzsche*, *Klages* summarized this experience of the world and its meaning for man in the sentence: " the image which impinges on the senses, that and that alone is the meaning of the world". Since all the forms (Gestalten) and images of the world - the sky, the countryside, the people, the animals and plants - undergo incessant change and therefore appear to be living, it is only logical for the natural philosophers to conclude that not only part of nature, but the whole of it and the cosmos is living and also animate, that there is no basic difference between non-living and living matter. The natural philosophers hence classify the spheres of the world in degrees of their perfection. The universality of life in the cosmos leads them to a holistic image of the universe in which all domains are meaningfully interconnected. They are content to contemplate the universe and remain amazed at the mystery of life. This explains their reluctance to intervene in nature, to analyze it or to carry out experiments.

2.2. The Natural Scientific or Reductionist (Teilinhaltliche) Theory of Life

Comparison of the natural philosophical with the natural scientific theory of life makes the difference between them quite clear. Unlike the former, natural science does not seek images which it can order and classify, but enquires into the causes and mechanisms of living processes; it directs its questions to the "whither" and "how" of these processes. *Klages* therefore termed natural scientific investigation of life "research into causes". The objects of this science are not the phenomena of life itself, but their abstractions: the conceptional things, the facts, the causes, the forces, the times [though these can only be considered (abstractly) on the basis of their experienced figurative contents].

The representation of a thing ("Ding an sich") is the existent nature of the Ego which is projected into the world; representation of the capacity of things to exert an effect, of "cause" and "force", is the self-assertive character of the Ego (*Klages*). Even *Nietzsche* had realized that the concept of mechanical action stems from ourselves: "the extraordinary persistance of belief in causality causes us to believe that everything which happens is an action" (and elsewhere) "the only power which exists has the same kind as the will". Thus, anyone seeking causes projects his own will on the objects investigated.

If such an approach dissects nature into acting things, then of necessity it is also split up into units (or elements); qualities, i.e. perceptible attributes, are converted into quantities (or numerical parameters). Hence in a strictly scientific sense something can only be conceived if it has an attribute which can be expressed numerically. This explains the natural scientist's obsession for expressing all his results in numerical terms (see Chapter VI, p. 147).

The methods of the natural scientist include conceptualization of things and facts as well as their causal association (which is termed causal inference and constitutes the substance of logic-rational thought). This causal-rational thought is not concerned with living, polar and meaningful connections; its aim is to establish causal relationships, possibly represented in mathematical form.

Bodily processes are resolved into subprocesses which are analyzed by physical or chemical methods and their causal connections determined. There can be no gaps in a causal chain because the relationship of cause and effect would then be lost. Moreover, in consistent natural scientific explanation, a "psychic factor" cannot be introduced at any point in the chain of causes because the physical or chemical laws would thereby be broken. A psychic factor such as the Vitalists termed "life force" could not be introduced into the "law of conservation of energy". In reality, this "life force" is not an externally acting force in a physical sense, but a *meta*physical agency effecting changes from inside and which consequently cannot be interpreted in energetic terms. True transformation characterizes everything that is alive - as *Klages* correctly observed. However, according to the laws of classical physics such a transformation is impossible.

Purely scientific research into life does not yield insights such as those from phenomenology (Wesenslehre), but merely data which serve to establish abstract conceptual systems and laws. For the natural scientist, the world can be divided into many different compartments. An unavoidable result of this abstract thinking is separation of living and non-living things. In strictly scientific thinking, living and non-living are no longer grouped (as by the natural philosopher) according to their degree of beauty, perfection or sublimity, but are sharply separated from one another.

The ultimate aim of the contemplating natural philosopher is the respectful adoration of the mystery of the world, whose meaning he seeks to experience and to understand. The natural scientist tries to "get to the bottom of" and to control the processes of nature and life.

Some examples of natural philosophical and natural scientific thinking are set out in Table 1, though I do not discuss them in detail here.

2.3. Limitations of the Two Lines of Thought

Where are the limits of the two lines of thought? As shown by the history of the last century, natural philosophy (which is able to form a picture of the world which is self-sufficient) has fallen into discredit because of a few inferior adherents who not only tried to order the phenomena of life according to polarities and analogies, but also attempted to include "things", "facts" and "forces" in their sphere of thinking. The danger began as not only manifest life, i.e. directly perceptible living phenomena, but also non-manifest life, i.e. those "phenomena" which are only seen after dismemberment of an organism, was drawn into the domain of natural philosophical consideration and interpretation. If natural philosophical consideration is restricted exclusively to a closed circle of meaningful relationships, i.e. to irreducible wholes (Ganzheiten), it can lead to highly meaningful and valuable results. However, a mixture of natural philosophical with natural scientific thinking and procedures must be repudiated. Research into entities and research into causes are mutually exclusive; they open up two different worlds. They can yield complementary information which may clarify a particular problem though their methods and aims must always be kept separate.

There are also limitations to natural scientific thinking. Although natural science appears to be unrestricted within the domain of its methodology, the natural laws and rules which it establishes are always valid only under quite specific conditions, a circumstance which limits their unrestricted applicability.

This can be illustrated by an important question about information theory which has been much discussed recently. The information theorists *N. Wiener*, *W.M. Elsasser*, *W. Wieser* and *J. v. Neumann* have been discussing whether the same natural laws are applicable to an organism as can be applied to a robot. According to these authors, any task which can be formulated logically can certainly be carried out by a robot, irrespective whether by physical or chemical means. It is particularly significant that (as demonstrated by *von Neumann*) robots can be designed which can repair or indeed reproduce themselves. Naturally the technology required would be so large-scale that there is hardly a celestial body large enough to accomodate it. However, it is quite possible to imitate any biological behavior with a robot provided that the behavior can be logically formulated, i.e. interpreted causally. A robot only yields as much information as assumptions (i.e. programs) were fed in; it can never produce its own programs or replace information which has been lost.

Experimental investigations have shown that information storage (and also scanning of these data) occurs in a space of microscopic dimensions in the organism. A reductionist (teilinhaltliche) biology (i.e. one orientated towards natural science) would encounter serious barriers if it tried e.g. to derive all the morphological and physiological data of an organism from measurements on the germ cell from which it has developed. If ontogeny, i.e. the development of an individual living creature, were an automatically proceeding sequence of events, such a prediction should be possible even if nothing were otherwise known about the organism concerned.

Investigation of the structural properties of the genes shows that only a few amino acids occur in the macromo-

lecules (the chemical building blocks of the genes). With
characteristic repetition, they are strung together in
a sequence of several hundred thousand amino acids. Assuming for the sake of simple calculation that only 10
amino acids can act as building blocks and that a protein
molecule consists of 100,000 amino acids, there are
$10^{100,000}$ different sequences of this length. It can moreover be calculated that there have not been more than
10^{66} species of organisms in all known planetary systems
from the beginning of the world until today (*R. Repges*).
If one wished to scrutinize every possible sequence of
amino acids with respect to its morphological and functional potential, then there is a maximum of 10^{66} available. It is quite impossible to determine the probable
potential of the other immensely numerous possibilities,
quite apart from the experimental difficulties.

These considerations lead to the general conclusion that
a reductionist (i.e. causal) interpretation of vital
processes must produce increasingly uncertain results
the nearer one comes to the molecular level. It is nevertheless known that vital processes show a high degree
of order throughout the living world, as shown by the
inheritance of certain organismic peculiarities over many
generations.

The most important feature of heredity is the transmission
of an image (symbolic representation). This mysterious
life process can only be grasped by figurative, holistic
thinking. This example may be taken to substantiate the
limitations of the causality and logic of reductionist
thinking and to illustrate the importance of a holistic
consideration of vital processes. A further example is
afforded by reflex theory.

The physical concept of the reflex and reflex movements
is first known to have been used in scientific considera-

tions of vital phenomena with the rise of rationalism in the seventeenth century. Before this time, body movements had been mentioned (but also mind and spirit movements) though not "reflex movement". The reasoning which had already subdued objects which were incapable of spontaneous movements (in contrast to animals and humans) to a mechanical law now also attempted to place the movements of animals and men under the laws of mechanics. It is hence highly instructive to follow the steps with which this process of "mechanization of life" was pushed forward almost by force.

In his book "De Homine", *Descartes* was the first to try to explain with the aid of a purely mechanistic explanation all of the processes in the human body. (Incidentally, *Descartes* did not publish this book in his lifetime for fear of the Inquisition; it was first published after his death). According to *Descartes's* view, the human body is a machine which unwinds like a clock or a perfect automaton in accordance with the laws of its driving mechanism. He studied the "mechanism" of various organ functions in detail. Certain living processes (such as the eye-lid reflex) are discussed in detail and - this is remarkable - compared with physical reflexes. *Descartes* thus applied the physical concept of the reflex to a physiological process for the first time.

I have given an exhaustive history of reflex theory elsewhere (*Blasius*, 1965). Only the modern interpretation of the reflex is of interest in the present context.

Since the inception of a detailed electrophysiological elucidation of the action of the parts of the reflex arc (particularly of the muscle spindle and tendon receptors), the concept of a "regulatory system with feedback" for muscular activity (in which the reflex is incorporated) has won more and more recognition. This model borrowed from engineering is able to show a satisfactory causal relationship between numerous facts.

However, this does not account for the meaning of a reflex; the reflex process must be contrasted with natural movement.

If we look at a lively movement and ask ourselves what part of it could be interpreted as reflexive, then we observe that the natural movements - especially those expressive movements which are particularly lively, graceful and beaming - lack exactly those attributes which we find in reflexes: mechanical, automatic and calculable characteristics. Thus, we would have to admit that the livelier a movement, the less automatic it will appear and the less are the chances for it to be calculable and predictable. If all movements were completely reflex in nature, spontaneity would be excluded.

All the reflex movements which can be demonstrated experimentally by physiological methods contrast with truly living movements, which are characterized by spontaneity. Reflex movements reside exclusively in the somatic domain, are interpretable in causal and mechanistic terms and are hence predictable.

The following observations can be made on the epistemological aspects of the problem. If a living organism is considered as a whole, this entity, form (Gestalt) and image can only be comprehended and described in qualitative terms. The psychic attributes revealed by the movements of expression can only be perceived at a definite time and place, i.e. uniquely and spontaneously. The description of unique events which can not be repeated is the concern of psychology - in the best sense of the word -

and the concern of the study of entity, form (Gestalt) and expression.

On the other hand, if the physical functions (i.e. functional details) of an organism are to be understood, then it is entirely legitimate to carry out experiments on this organism so far as this is feasible. In this respect the triggering of a reflex is always an experiment which shows quite definite results. Stimulus and excitation can be interpreted according to physical laws.

However, this problem of research into life has also its psychological side. The polarity of stimulus and excitation and the meaning of movement should be dealt with in terms of psychology.

These examples from information theory and reflex theory have been selected to show the limits of a natural scientific treatment of living processes. I should now like to give some instances where the limits have been exceeded

There is an important danger of natural scientific reasoning and procedures, particularly since its methods are not restricted to the material and energetic aspects of natural processes, but are also applied to their psychic side, i.e. the essence and meaning of living phenomena. This exceeding of bounds is just as inadmissable and dangerous as is the encroachment of natural philosophy in the domain of natural science. "In the world of things, causes and forces, there is no room for souls; the home of souls is the reality of images" (*Klages*). True research into entities and logical research into causes are hence concerned with two entirely separate domains. The greatest danger for natural science is that at bottom it has no final objective because it is unbounded; every cause has an endless chain of causes, leading to a "regressus in infinitum". This is also the reason for the perpetual dissatisfaction of the natural scientist (in the strict sense); all his efforts are basically aimless and hence in vain. Besides this, the striving of the natural scientist to control natural processes and subdue the world leads to a disintegration of images and collapse of the image of the world. This disintegration began as soon as the thought on life and nature postulated a separation of living and non-living matter. *Heraclitus* could still say: παντα πετ i.e. "everything without exception is alive". In the natural philosophical sense, the whole universe appears as a living whole. *Klages* expressed it thus: "Without exception, all images, phenomena or entities are living, while all abstract things are non-living. Conceived as perceptible images, not only plants, animals and humans are living, but also rocks, clouds, water, wind and flame, the sky, the earth and even space and time. Conceived in the sense of an object which can only be thought, man (just like any other object) is merely an agglomeration of atoms in mechanical movement".

Life is thus outside the world of abstract things, facts and causes and can only be found in the real events of the images. The shaping force of images should not be viewed as a physical force which is movement influenced from the exterior, but as a metaphysical effectuation which changes from inside. "True transformations are impossible in terms of physics, but incessant change characterizes every living thing" *(Klages)*.

It will never be possible to reassemble into a living, complete whole the elements of a living organism dispersed by experiment and analysis. In view of the depressing proliferation of individual facts furnished by abstractive thinking in the development of a theory of life orientated towards natural science, we have to ask whether it is possible at all to form an integrated whole for the science of life from this abundance of data. This question has to be answered in the negative. Abstractive thinking is so structured that it can never conceive more than parts of a whole. Only the contemplating soul is able to comprehend an irreducible whole which is not more than the sum of its parts but also something entirely different. Therefore, if biology wishes to outline a global theory of life, it must accept the doctrine of natural philosophy and even promote it; for the important reason that only the latter is able to give insights into the essential nature of life.

There are two ways of observing nature. One is holistic-psychological, attempting to interpret the meaning of the phenomena of life. The second is reductionist-natural scientific, yielding a causal interpretation of somatic processes. Mixing the two approaches leads to paradoxes such as those which I have attempted to illustrate.

How should we deal with this epistemological dilemma? I believe one should train oneself to be able to adopt both approaches in order to exhaust human cognitive potential fully. We should hence acquaint ourselves with the essence, value and limitations of the two possible theories and beware of excessive emphasis on the one which is more attractive to our particular personalities. *Johannes Müller* represented this view which is proposed in his doctoral thesis at 21 years of age in 1822: "psychologus nemo, nisi physiologus". In the present situation in research into life, this should be reversed: "nobody can be a physiologist without being a psychologist".

3. The Scope of the Natural Scientific Theory of Life

The theory of the forms and functions of living organisms (in the widest sense) is today the domain of the science of biology, while "physiology" merely entails the theory of organismic functions. Biology is hence subdivided into two huge fields: *morphology*, the theory of the form (Gestalt) of living organisms[1], and *physiology*, the theory of living organisms' function. Physiology is further subdivided according to the object of study into vegetative physiology, which is concerned with storage of energy and with assimilatory processes, and animal physiology, which is concerned with the discharge of energy and with dissimilatory processes. Vegetative physiology includes digestion, respiration, cardiovascular system, blood, metabolism and excretion. Animal physiology includes the tasks of the sense organs, nerves, muscles and the central nervous system. Today physiology is usually subdivided according to the methods employed to study the vital processes; hence physical physiology is distinguished from chemical physiology.

The entire field of physiology can be systematically subdivided into general, special, comparative and applied physiology. "General Physiology" describes the physiological principles which apply to all organisms and which are characteristic for all organs, particularly those of man and higher animals. These principles include the chronological laws of organisms, the laws of energy, the laws of excitation and adaptation etc. The vegetative and animal functions are treated in "Special Physiology". The

---

[1] Scientific morphology is no longer a theory of form in the sense of natural philosophy; it does not interpret, but merely describes the forms of life.

task of "Comparative Physiology" consists of comparing the functions of particular organs in different animals and of describing and analyzing from a comparative standpoint the abundant possibilities of distinct physiological needs. In recent decades, "Applied Physiology" has attained an ever increasing importance since it was possible to apply successfully physiological principles in pathology, in the hospital, in sanitation and in engineering. Independent research fields have thus come into being for pathological and clinical physiology. Favorable conditions of work, limits of working capacity and the conditions of recuperation and fatigue are studied in industrial physiology and sport physiology. In the field of high altitude physiology established during the last war the conditions of living at great altitudes, questions of artificial oxygen supply and adaptation to altitude are explored. Finally, the recently developed space physiology is concerned with the effect of gravity on the human organism and the problem of weightlessness in space.

3.1. Methodological Foundations of Physiology as Natural Science

As an empirical scientific theory of the vital functions of man and the higher animals, physiology is concerned with observing natural processes, their experimental analysis and with stating their principles.

As a branch of experimental science physiological research has recourse to the methods and principles of physics and chemistry. Since physics and chemistry can be combined in atomic theory, all branches of natural science (and thus also physiology) are consistent with the absolute system of measures. A prerequisite for this general system of measurement is the existence of fixed standards (units of measurement, universal constants) and mechanically invariable clocks (time units). One could object here that units of space and time can never be constant: the motion of the earth around the sun varies in its temporal duration [2]; even the metric units are fixed arbitrarily and do not correspond to any absolutely constant correlates in nature. For the present of course these objections are not yet of any concern for practical work in engineering or for physiological methodology.

3.2. Natural Laws and Rules

A single measurement is of no interest to the natural scientist. His objective is to state rules and laws which can be derived from a very large number of single observations and measurements. Physiology also endeavors to establish a system of exact laws for the organismic domain such as that which already exists for the extra-organismic domains in physics and chemistry. Physiological observations can thus be designated "exact" in the same sense as the physicist uses this word.

The discovery of exact laws is going on in many fields of biology, particularly that of physiology. One can mention as examples the general energetics of organisms, the quantitative laws of metabolism, the laws of growth, the chronological laws to which organisms are subject, the theory of excitation and nervous conduction, the quantitative laws governing the activity of sense organs, the laws of reflexes, of nervous centers and the influence of certain

---

[2] Even the most precise instruments for measuring time have an error factor which can change depending on the conditions of the experiments.

substances on the organism (e.g. the hormones and narcotics), the laws of heredity and much else.

The discovery of laws which permit a "unitary" consideration of diverse phenomena leads inevitably to a kind of "homogenization of reality" (*von Bertalanffy*). It was difficult for people in the 17th century to imagine that phenomena so qualitatively different as the orbit of the planets, the free fall of a stone, the upward striving of a flame, the movement of a body on an oblique plane, the swinging of a pendulum, etc. could be subsumed by *one* law of motion. Mechanics and thermodynamics were similarly combined later in the mechanical equivalent of heat, as were optics and electrical theory in the wave theory. In this fashion still "more uniform" laws were revealed.

A similar process is taking place in biology and physiology. A prerequisite for discovery of such laws which can summarize life processes in a uniform fashion is the measurement of these processes. Even physiologists follow the guiding principle of *Galileo*: "measure what is measurable and render measurable that which cannot yet be measured".

3.3. Limits of a Metrical and Mathematical Treatment of the Phenomena of Life

Any measurement is based on an objectivization, on a process of abstraction which extracts a homogenous body of facts from the abundance of phenomena available. Innumerable properties of the organism can only be contemplated, described, felt. Here natural philosophical research comes into its own. In this field, descriptive, morphological natural science unfolds itself. Form and image, however, are also to a certain extent measurable. An exact method can clarify certain problems of form. Mathematical symbols are related by rules of calculation which have a logical foundation; opinions and interpretations are excluded here. There are in essence only two possibilities with respect to an exact theory: to accept it, if it is adequately proved, or to demonstrate its falseness by facts which are quantitatively verified.

It is often asked whether quantitative relations are not too abstract and schematic to apprehend living reality. This distrust of exact procedure is often met with in individuals interested in biology who belong to the cyclothymic type which thinks in plastic and concrete terms and dislikes the abstract (according to *Kretschmer*). However, descriptive, intuitive description can be of value in science as a counterbalance to the exact logical comprehension of quantitative laws with which the schizothymic type is particularly concerned [3].

---

[3] *Kretschmer*'s constitutional types are classified as leptosomic, pyknic and athletic according to physique on the one hand and as schizothymic, cyclothymic and ixothymic according to temperament on the other.

The leptosomic temperament alternates between poles of hyperestheticism and non-estheticism. The leptosomic is designated a schizothymic because of his faculty of splitting off his inner world from the exterior. Intensive analytical thinking, rational association and abstract expression suits the leptosomic and he has a propensity for formalism and systematizing.

The pyknic shifts in the change of his feelings between extremes of elation and melancholy (hypomanic-depressive). The synchrony of his inner and outer world characterizes him as a cyclothymic; his thinking is synthetic-extensive, i.e. he tends to apprehend "Ganzheiten" and the diversity of living phenomena; his mode of expression is matter-of-fact and figurative. He is hence a vividly describing and probing empiricist.

The athletic personality is characterized by an alternation between a dynamic-explosive and a viscous-enechetic disposition. The basic feature of his being is the tenacity and ponderousness of his inner world as it relates to his outer world; he is hence designated an ixothymic (ιξος = bird's lime). The stability of his temperament gives him strength of character and peace of mind in difficult situations; he has a cautious and straightforward mode of thinking.

3.4. Mathematical Methods

Mathematical methods are an indispensable tool in many areas of biology and physiology. Statistics are used to draw up specific rules which serve to calculate the average and range of the relationship between two or more functional quantities. For example, investigations on a large number of healthy subjects of the same sex are necessary to show the correlation of heart minute volume with age; the resting minute volumes are measured for different ages. In other cases, correlations which can be represented by a mathematical function (e.g. growth-rule) can be determined. It is not necessary, however, to calculate the correlation coefficient to more decimal places than can be put to good use.

However, mathematical methods of analysis are necessary to gain insights into physiological principles. Basically they consist of certain hypotheses about basic facts of natural events. These are formulated mathematically and inferences are made which can be tested by experience and experiment; if they pass the test, then the hypothesis is confirmed. The analytical process is based in the final analysis on differential equations which enable relationships to be formulated. If the relationships investigated are verified, then they are natural laws, i.e. rational relationships which do not describe the phenomena, but explain them in terms of general principles. In applying this method, reason "triumphs" to some extent over intuition, as is also the case in physics. The "law of conservation of energy" described by *Robert Mayer* 1842 in Liebig's Annals can be taken as an example for a general principle which is valid not only for all physical and chemical processes in closed systems but of a general universe if it is also viewed as a closed system (cf. also the expositions under 5.1.).

Statistical rules have been distinguished from natural laws. Surely it should be mentioned that natural laws are also derived by statistical means. Like statistical rules, natural laws are basically of statistical character, i.e. they constitute average values of numerous events which cannot be considered individually or are indeed too numerous to be registered.

3.5. Parameters

Any statistical rule or natural law is only valid under certain conditions which must be established by experimental testing. This can be illustrated by an example: in order to specify the flow resistance W in a narrow tube (e.g. a capillary) through which a fluid is passing, the law discovered simultaneously by the German engineer Gotthilf *Hagen* and the French physician Jean Louis *Poiseuille* in 1839 is applied:

$$W = \frac{\eta \cdot l}{q^2} \cdot k \qquad (1)$$

where η equals the viscosity and l = the length of the tube. If the dependence of W on q is represented as a power function, then η, l and the constant (k) are to be designated and treated as conditions of the equation. However, the equation derived from the physical model only applies under the further conditions that the wall is wettable, that flow is laminar, that the fluid is homogeneous and that the tube has a rigid lumen. The fact that these conditions can never be kept the same in the living organism limits the applicability of the formula to the organism.

3.6. Validity of Laws

Natural laws apply only under constant internal and external conditions. If the conditions or parameters change, natural laws must be formulated in a different way. Scientific progress consists in knowing all the conditions of a law as precisely as possible and in analyzing them experimentally. If it becomes known that a process does not occur randomly but according to definite rules or laws, then there is a possibility of controlling it or making predictions about the course of the same or a similar natural process. If the conditions of the process are within the limits of what can be analyzed by the law, then it is a matter of an extrapolatory prediction (usually with a low probability of being true).

4. Special Attributes of Organisms

The distinction between living and non-living nature (which was still much discussed in the last century) is today better considered as the difference between organismic and extra-organismic nature. There are no basic differences between forms, materials, forces and functions in these two spheres of nature. However, organismic matter has certain special features and another kind of order which have to be interpreted in a holistic, figurative sense. It is usually easy to decide whether a structure is organismic by reference to certain characteristics; any or all of these special properties are unlikely to be found in an extra-organismic system. The most important organismic characteristics are enumerated below:

4.1. Cellular Organization

As a whole, organisms can be distinguished as entities with cellular structure; extra-organismic nature does not have cellular structure. The cells of an organism are made up of protoplasm. Protoplasm is not to be viewed as a chemical substance or as a physical system, but as a complicated structural entity which finds itself in continuous internal change manifested externally as "flow".

A link between cellular and non-cellular structures has been found recently in viruses responsible for certain diseases and also in certain viruses which are important for the health of the organism. Such viruses do not have a cellular structure, but rather consist of single molecules or molecular aggregates (which can even crystallize under certain conditions). Although viruses do not have a cellular structure, they are capable of reproduction. This is not possible outside a cell, but within a cell they suddenly proliferate by "identical reduplication" in which the substance for multiplication is supplied by the host cell and the virus only provides the matrix. It is interesting, for example, that the rabies virus (which causes the affected animal to be afraid of water and to bite) creates favorable conditions for multiplication of the virus.

4.2. Chemical Constitution of Organisms

All organismic substance has a specific constitution. Only certain chemical substances are present in the living cell, except water, essentially proteins, fats and carbohydrates. All these substances contain carbon, which obviously has a special role in the functioning of the living organism. Carbon has the special property of being able to bind both positively and negatively charged atoms, e.g. oxygen and hydrogen respectively. Carbon is also able to form bonds with itself, giving rise to rings and chains; this explains the enormous diversity of carbon compounds. However, it must be remembered that the availability of inorganic substances such as O_2 and CO_2, the elements magnesium, sodium, potassium, calcium etc. are essential preconditions for the life of plants and animals Traces of these elements are connected in particular with the excitability of cells. Unprejudiced consideration enables one to state that the organismic and extra-organismic world are in dissolubly linked and that they permeate each other.

4.3. Transformation of Energy

Life can be seen as a continuous transformation of energy and the organism viewed as the transit point of a continuous energy stream. The organism obtains potential energy from its food, transforms it into specific energy and releases it again as kinetic energy of the vital processes. Such a system can be termed an "open system" because it is maintained by a "steady state" (cf. 5.2. ff.). Just as a flame has the same shape and height, although continuously fed with new combustible material and oxygen, or like the water in a fountain which is renewed over and over again and yet retains shape, so the unique manifestation of a living organism does not depend upon the molecules of the substances of which it is composed. The basic substances are subject to a continuous relatively rapid replacement like those of the flame; this replacement or "turnover" of substances is closely related to the vital activity of individual organs. Some constituents of the body are renewed after days, some after months and others only after years. As in the case of the flame or the

fountain, what remains as the individual organism is the
form or image which incessantly takes new material into
itself and loses old material. The kind and extent of the
transformations of energy are species - specific and inherent. The laws of energy, i.e. the principles of thermodynamics, will be discussed in more detail under the description of open systems (cf. 5.2.).

4.4. Relations with the Environment

Every living organism stands in a quite special connection
with its environment which is to be interpreted as a polarity in the natural philosophical or psychological sense.
Seen from a natural scientific point of view, the extraorganismic world is only part of the whole universe; and
is in reality interrelated with every organism. One can
indeed say that a living organism has a polar connection
with the whole cosmos. The human being is informed about
the world by the behavior of solar radiation to the earth
acting on the eye, which is equipped for exclusive uptake
of this radiation. The ear would not be able to function
in the absence of the terrestrial air envelope. The otolith apparatus is attuned to the gravitational field of
the earth. Respiration requires the oxygen of the air;
the organs of digestion are organized specifically to
process vegetable and animal food. Human organization has
thus evolved precisely in line with the exigencies of
the world in the same way as social and sexual behavior
is attuned to the social environment.

All these realities cannot be comprehended in causal terms,
but must be accepted as inherent necessities of nature.
The question arises as to what are favorable conditions
for human life. Every creature - therefore the human individual too- only has an environment conducive to its fur-

ther life. One creature seeks out congenial and another hard conditions of life; these serve and suit its fulfillment. Hard conditions of life promote performance in many people and increase their attainments; the capacities of many individuals increase in face of resistance and adversity. There is a polar relationship between action and reaction. This exigency of nature and its importance for human beings was aptly expressed by *Goethe*: "the master first shows himself when constrained".

There is an essential difference between organismic and extra-organismic nature in the following respect: organismic life creates disequilibria in certain phases or resolves itself into polarities; its energy situation tends towards a state of minimal entropy, i.e. minimal disorder. On the other hand, extra-organismic nature provides for an equilibration of energies, a state of maximal entropy, i.e. maximal disorder; of course this statement only applies to closed extra-organismic systems. The conditions of the general energy laws applying to open systems are different in nature and are discussed later.

4.5. Evolution of Life

A peculiarity of all living things is their phylogenetic evolution. This evolution can be traced from unicellular via multicellular organisms to the higher forms of life. The most noticeable feature in phylogeny is the ever more pronounced differentiation, specialization and division of functions. This development does not only entail the self-assertiveness of organisms or the "struggle for existence" (*Malthus* and *Charles Darwin*); there are deeper reasons. The multifariousness of the forms and characters, the struggle for self-expressiveness, the joy in life or the development of personality are not touched at all by this principle of "scientific" explanation. The *Darwinian* principle does not explain why life always arises a new. The continuous stream of life is thoroughly mysterious and cannot be explained in terms of natural science.

Independence from the environment has increased in the course of evolution from the unicell to the highly organized organism. This development in the plant and animal kingdoms and in man hence tends to make the organism as independent as possible of the external conditions, i.e. attain a greater degree of freedom. (E.g. the plant is restricted to a particular location, while animals are mobile). The functioning and metabolism of a cold-blooded animal depends on the environmental temperature, while appropriate regulations make the warm-blooded animal independent of its environment. However, independence of the environment is won at the price of greater insecurity. The mobility of the well-armored tortoise is highly restricted, though this animal is difficult to attack and reaches a great age; on the other hand, the bird is little protected though it can move in any dimension of space; it can be injured very easily and has a very short life. Man, too, has become very much more mobile with modern means of transportation and has thus gained a higher degree of freedom, though this has been gained with a greater risk to life. The question of the perfectibility within the whole realm of living creatures is difficult to answer. Which is the more perfect, a plant such as a rose, or an animal such as a horse? Both are perfect in their own way. Is man perfect? Though man is the most concerned about perfection, human powers of comprehension and especially the enormous efforts of human will power tend to suggest that man is finally destined to obliterate all terrestrial life and to be the catalyst of the destruction of the earth.

4.6. The Course of Life

The course of individual life, ontogeny, is a further characteristic of life. Life always arises as an unicell, generally by reproduction. Life does not begin with a jerk, but unfolds as previously existing cells become independent. These special characteristics also determine the laws of heredity and growth. The extent of growth and all essential characteristics of an organism are inherited as are the cessation of development, ageing and transformation at death.

4.7. Excitability

A particular property of living creatures is excitability. If an external stimulus impinges on it an organism can react in a quite definite way. However, the intensity of a stimulus does not have any proportional relation to the kind and intensity of the reaction. This peculiarity contrasts with the laws of action and reaction in physics in which there is a strict proportionality. However, the organism is not a resting entity which is only stirred into action by external stimulation, but rather is a "system of internal activity" (*von Bertalanffy*), even under unchanging external conditions. The biological reactivity which is one of the special characteristics of every living entity is always manifested in the discharge of stored energy. Because recovery is a prerequisite for the activity of an organism, it is the most important process in the phase which follows the reaction. Then readiness to react is built up again. Every biological reaction constitutes a non-reversible process in which energy is always simultaneously released as heat.

The efficiency of an organism improves the more frequently it comes into contact with the same stimulus. For example, the efficiency of muscular work increases with systematic training, i.e. the entropy decreases (cf. Section 5.1., p.36). Such processes cannot be observed in physics or engineering; the efficiency of a machine becomes ever smaller with increasing use (i.e. wear) and entropy increases. Improvement caused by performance is observed not only in muscle but also in ganglion cells and other structures of the organism. Intellectual capacities can also be enlarged; attentiveness can be improved, linguistic expression, logical and systematic thinking can be raised to a higher level of effectiveness by practice.

4.8. Capacity for Regulation

In contrast to extra-organismic systems, an organism "moves of itself" (αωτικινητος according to *Alkmaion*); each movement is not the result of specific influences from the exterior. There is no process within the organism which proceeds against itself. The activity of life is directed to survival and self-fulfillment. The meaning and the importance of particular life processes is recognized in the search for functional cycles and for polarities. The polarities show the connection of all alive. If vital processes are disturbed, their smooth functioning is restored again. A recovery phase always follows a phase of activity; a state of hunger is compensated by uptake of food. The organism has an abundance of regulatory mechanisms for efficient compensation of disturbances (cf. 5.4.).

4.9. Animation

Last but not least animation is a characteristic common to all life; according to the natural philosophical view it is not only possessed by man but also by animals, plants and even the earth (as expressed by its physiognomy). As *Novalis* puts it, the seat of the soul is to be sought where the inner and outer world meet. This is the wide field of phenomenology with which natural science cannot deal, nor contribute any essential understanding.

5. The Flow of Energy as the Most Important Principle of Scientific Biology

Of the stated properties of living creatures, the continuous flow of energy (see 4.3.) is the most important and most general feature for the scientific interpretation of life processes. *Robert Mayer* developed the fundamental "Law of Conservation of Energy" to bring all the various forms of energy involved in vital processes under one general law.

5.1. The Energetics of Closed Physical Systems

According to *Robert Mayer*'s Law or the First Principle of Thermodynamics of *Hermann von Helmholtz*, the total energy of a closed system remains constant in all transformations; it can only appear in a different form. If such a system is subjected to an arbitrary change, the change in internal energy U is equal to the sum of the amount of heat released to the environment Q and the work A performed:

$$U = Q + A \qquad (2)$$

or described by an equation of differences:

$$\Delta U = \Delta Q + \Delta A \qquad (3)$$

The most important consequence of the First Principle for biology is the "Law of Constant Total Heat". Equation (2) only gives information on the energy content before and after a reaction has taken place and nothing about the route taken. The heat emission is always the same irrespective of the way in which one substance is converted into another. This law suggests the possibility that the heat changes in reactions which convert one substance into another (or the temperature coefficient) can be ascertained from the heat of combustion of the substances. This law constitutes the basis of the physiology of energy metabolism. Processes in organisms involve complicated stepwise reactions which cannot readily be followed individually. However, according to the "Law of Constant Total Heat", the amount of heat released from 1 g glucose on direct combustion to CO_2 and H_2O is the same as that released in the complicated reaction steps of the respiratory process. It is hence possible to apply the caloric value of food determined by direct combustion in the bomb calorimeter to the energy balance of the organism.

The budget of the energy transformations in the organism is determined from the caloric value of its food on the intake side and from the work performed plus the heat evolved and the heat of combustion of the excreta (all expressed in calories) on the expenditure side.

It is not possible to tell whether the heat evolved by an organism is "primary heat" arising in the resting state, or whether it is "secondary heat" from some kind of work which has been transformed into heat on the basis of the First Principle. Work which has been performed (A) could have arisen in two ways. Since in the final analysis all the energy transformed in the organism arises from the chemical energy of ingested substances, the chemical energy could have been converted into work in two ways: the organism could function either as a thermodynamic system, i.e. the chemical energy could first be converted to the heat of a higher temperature and this then transformed into work, or it could function as a chemodynamic system, i.e. the chemical energy could be directly converted into work (and heat). Which of the two possibilities is the right one cannot be deduced from the First Principle. However, the lack of substantial temperature differences in the organism shows that only the second possibility is correct.

During the conversion of heat into work (e.g. in a heat-powered machine), the Second Principle predicts that work is always transformed into heat, but only a part of the heat can be converted into work. The work performed (-A) by a heat-powered machine can be calculated from the formula which follows:

$$-A = Q \cdot \frac{T_1 - T_2}{T_1} \qquad (4)$$

where T_1 is the initial temperature, T_2 is the temperature of the condensation products and Q the added heat.

Formula (4) can also be written:
$$-A = Q \cdot \zeta \qquad (5)$$
where ζ is the efficiency or effective power of the heat-powered machine.

The "Second Principle" thus states that
1. heat never moves from a system of low temperature to one of higher temperature without performance of work; i.e. a temperature difference can never arise spontaneously;
2. in the conversion of heat into the energy of movement, into electrical energy, into radiation etc., part must always be given up as heat to a system of lower temperature;
3. conversely all these forms of energy can be converted entirely into heat.

The proportion of the inner energy which cannot be transformed into higher forms of energy is the entropy of the particular state (εντρεπειν = to turn in on itself). According to *Boltzmann*, entropy is the measure of the probability of the state in which a system finds itself. Increasing probability of a state entails an increasing disorder of the molecular constituents of the system. Any higher form of energy is a more probable state compared to the random movement of molecules which is manifested as heat. Transformation of all the energy of the universe into the form of heat, i.e. into a state of maximal entropy would eventually convert the universe from a more improbable to a more probable state. This would lead to a state of maximal entropy and hence to the "heat death of the universe". This gloomy prospect, however, results from an inadmissible simplification.

5.2. The Organism as an Open System

Considered as a whole, the organism has properties similar to those of equilibrium systems. In the cell and in a multicellular organism is found a specific composition, a constant relationship between constituents that resembles the distribution of components in a chemical equilibrium system. This biological equilibrium is largely independent from the absolute amount of components and is kept constant despite varying uptake of nutrients and changing external conditions. A disturbance in any part of the whole system elicits a particular change resulting in a return to the normal state; organic regulation is founded on this behavior. Any living organic system has a continuous metabolism which maintains a turnover of its constituents. The organism hence does not constitute a closed system, but an open system. A system is designated closed if neither material nor energy enter or leave it. An open system is one in which energy and material are imported and exported.

The basic property of a living system is simultaneous dissolution and regrowth, a linking of continuous degradation and synthesis. If degradation and synthesis are in balance, the living system appears stationary when viewed from the outside. Life can hence be termed a "bivalent (or polar) autonomous state of change". Every living creature is able to produce its own specific components in a characteristic kind and amount from a mixture of foreign substances (carbohydrates, fats, proteins, etc.) which would have a toxic action if absorbed directly.

To define this principle of incessant self-regulation of metabolism more precisely it is necessary to generalize the kinetics and thermodynamics to open systems. This is a problem with which *von Bertalanffy, Rashovoky, Hill, Prigogine* and others have recently been especially concerned. Such an interpretation has proved itself appropriate for a precise statement of the basic biological laws and also special processes such as organismic growth, stimulation, regulation, etc.

5.3. General Properties of Open Systems

"True equilibria" in closed and "stationary states" (named "steady states" by *Hill*) or "dynamic equilibria" (*von Bertalanffy*) in open systems show a certain similarity inasmuch as these systems remain constant when considered as a whole and with respect to their components. However, the physical situation is basically different in the two cases (cf. Table 2).

Table 2. General properties of closed and open systems

	Closed systems	Open systems
Equilibrium	Tending towards true equilibria	Constituting dynamic equilibria (steady states)
Reactions	Reversible, usually rapid reactions	Irreversible, usually slow reactions
Kinetics	Non-recurrent work	Continuous work
Thermodynamics	The Second Principle applies; the end state has a *minimum free energy* and *maximum entropy*	The Second Principle applies to open systems only considered together with their environment; constancy during turnover of elements; hence *maximum free energy* and *minimum entropy*.

True chemical equilibria in closed systems are based (1) on reversible reactions, are (2) a result of the Second Principle and are (3) characterized by a minimum of free energy.

On the other hand, the "dynamic equilibrium" in an open system is essentially irreversible since the products of the reaction leave the system. The usual version of the Second Principle only applies to closed systems, and not to dynamic equilibria.

According to the Second Principle, a closed system must finally pass into an equilibrium state which is independent of time (with maximum entropy and minimum free energy) and in which the correlation of the single phases remains constant.

On the other hand, an open system can stay in a state in which it can remain constant as a whole and in respect to its components despite changing elements.

Reactions in open systems cannot lead to a true equilibrium, but only to a dynamic equilibrium or steady state.

The system is maintained in this dynamic equilibrium by addition of material and energy and simultaneous loss of an equivalent amount of material and energy (as heat or work). The situation is comparable with that of a flame. The organism is only thus enabled to carry out life processes at all.

Coordination of the speeds of single reactions is important for the maintenance of the dynamic equilibrium in an open system. Rapid processes also lead to chemical equilibrium in the organism (e.g. equilibration of blood gases and hemoglobin). However, slow processes do not reach equilibrium but are maintained in a stationary state For example, the equilibrium state of sugar oxidation favors complete consumption of the sugar although the sugar level in the blood is kept constant because new sugar is continuously released into the blood from the food or from the glycogen, fat and protein depots. A certain slowness of the reactions is a precondition for the formation of a dynamic equilibrium; instantaneous reactions, such as ionic reactions, quickly lead to a true equilibrium. Maintenance of a dynamic equilibrium is only made possible by the fact that the organism is constructed from high-molecular-weight and complex carbon compounds. On the one hand these compounds are energy-rich but chemically inert so that maintenance of a large chemical potential is guaranteed. On the other hand, the enzymes enable a rapid and regulated release of these enormous quantities of energy.

A relatively high temperature of about $37^{\circ}C$ is maintained to increase the constant flow of energy and material in the organism. The temperature of protein coagulation sets an upper limit for body temperature. A more rapid turnover also entails a quicker re-supply to the system; this is the reason for the ever greater development of the organs of locomotion and circulation in the more differentiated animals.

5.4. Kinetics of Open Systems

The simplest form of transformation in an open system is a monomolecular reaction $a \rightleftarrows b$ in which reaction material a is continuously added while the products b are removed from the system by suitable mechanisms (e.g. diffusion, precipitation, etc.). The more completely the reaction products b are removed from the system, the greater the working capacity of the system. This is the case in absorption from the intestine for example; the greater the absorption capacity of the intestinal epithelium, the more rapidly digestion proceeds. The constants of the system, i.e. the reaction rate, for example, determine the relationship of the components in the reaction, though not the size of the influx. This is the basis of the principle of self-steering and autoregulation which ensures the relationship of the components to be kept the same despite varying influx and efflux from the system. If the speed of a degradation process is changed by a disturbance or a stimulus, a new stationary state is produced; the system attempts to compensate for the increased turnover by intensified uptake of reactants.

The characteristics of autoregulation are thus consequences of the general properties of open systems, i.e.:
1. Maintenance of a constant dynamic equilibrium with constant turnover of the components,
2. Composition is largely independent of influx,
3. Restoration of the dynamic equilibrium after an increased turnover caused by a disturbance.

Examples of dynamic equilibria: bioelectric potentials which disappear at death, differences in ion concentrations maintained across membranes, or reactions to stimuli (which amount to a change in one or several rate constants in an open system).

An important consequence of the theory of dynamic equilibrium as opposed to true equilibrium is that certain processes can be balanced by an overshoot compensation. Examples of this are: the positive after-potential which follows the negative after-potential of a nerve, the discharge after an inhibition, the hyperventilation after an apnea, the intensified metabolic activity after an oxygen debt, etc. From the point of view of classical kinetics these are paradoxical compensations and only occur in open systems with a dynamic equilibrium.

5.5. The Thermodynamics of Open Systems

Thermodynamic consideration of open systems gives results just as interesting as those from the kinetic consideration applied up to now. *Prigogine* commented in 1947: "classical thermodynamics is an admirable but fragmentary theory. It is fragmentary because it can be applied only to equilibrium states in closed systems. One should therefore attempt to found a more general theory which covers states of disequilibrium as well as equilibria". A mathematical presentation of the more general theory given by *Prigogine* and others can be dispensed with here, though the consequences of his theory must be described briefly.

Entropy must increase in all irreversible processes in closed systems; the entropy changes in these systems must hence be positive. On the other hand, in an open system (particularly in living organisms) entropy does not only increase as a result of irreversible processes, but the organism also "feeds on negative entropy" (*Schrödinger*), in that it takes up complex organic molecules, uses their free energy content in order to construct systems of higher order inside itself and returns the end products to the environment; in this way the organism decreases its entropy.

Living systems can thus maintain themselves in dynamic equilibrium by uptake of material rich in free energy and thus avoid the increase in entropy which is inevitable in closed systems. The validity of the Second Principle is not infringed by this formulation; it applies to the "open system with its environment" in that the free energy taken from the environment is used to keep the entropy in the system constant. The free energy is thereby to some extent "devalued", so the entropy of the whole system increases. The Second Principle is thus inappropriate to interpret adequately the state and transformations of the energy in an open system. If an open system is in a stationary state, the change in entropy is zero because the positive entropy resulting from the irreversible processes is balanced by the negative entropy resulting from input of free energy in the form of food.

In summary, it can be said that the following generally valid principles apply to open systems (*von Bertalanffy*):
1. Open systems approximate a stationary state with minimum of entropy; the entropy within the system may decrease in the establishment of such a state (giving the possibility of bodily growth).
2. These states with minimal or no change in entropy generally exhibit stability.
3. A change in one of the variables of state is counteracted by the system itself (giving the possibility of regulating organic processes).
4. In complete harmony with the laws of thermodynamics, open systems can pass spontaneously to a higher level

of organization, i.e. evolve a state of greater heterogeneity and complexity. The entropy can decrease in the formation of such a state.

5.6. Provenance of the Energy for Living Processes

Before one can discuss the source of the energy for vital processes, it is necessary to consider in detail the exchange of substances in nature as a whole. The findings of *Robert Mayer* and the studies of *Justus von Liebig* show that an energy cycle can be observed in nature; this can be represented diagrammatically as in *Figure 1*.

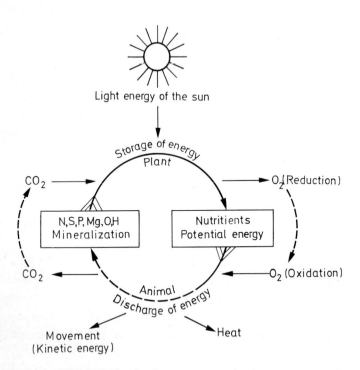

Fig. 1. Energy cycle in living processes (*Blasius*, 1962)

Under the influence of light, plants are able to synthetize certain nutrients by taking up carbon dioxide from the air with the aid of certain elements absorbed from the earth and by releasing oxygen. The potential energy arisin in this process derives ultimately from the huge energy re serve of the sun. Plant nutrients serve animals and men as food. The life work of plants consists essentially in discharging energy. Release of energy in the animal organism is only possible with simultaneous uptake of oxygen. One-thirtieth of the carbon dioxide absorbed by the plant is released again by the plant; the rest is used to synthetize plant carbon compounds of high molecular weight. It is peculiar that all the carbon in organismic nature derives from the very low carbon dioxide content of the air. The total reserve of carbon in the earth's atmosphere is about 21 000 million Kg and this is all exchanged between the atmosphere and animals and plants in about a year. It is just as significant that the bulk of the molecular oxygen in the earth's atmosphere derives from plant metabolism, so that in the history of the earth the plants created the conditions for animal and human life. Within the animal organism, there is not only discharge of energy, but a rhythmic alternation with energy storage. While the latter process constitutes the task of vegetative physiology, animal physiology is chiefly concerned with the study of discharge processes, i.e. with the interactions of the organism with its environment.

All the energy on earth arises from nuclear disintegrations taking place within the sun. It has been calculated that this energy reserve in the sun will last for several million years. Only one 23-millionth part of the energy sent out from the sun reaches the earth. Of this energy, only one eight-thousandth is used by all plants in the world. Man uses one three-millionth part of the energy of the plants.

As made abundantly clear by the data in the above paragraph, the life of plants, animals and humans is linked to the sun. As a living organism, man is indissolubly bound to all the other life on earth and to the sun; he is isolated only as a rational being. The uniqueness of man is only justified with respect to his spirit; he is otherwise part of the universal life of the cosmos.

6. Summary

If we return to the point of departure of our epistemological discussion on the essence and object of biological research, we come to the realization and knowledge that both a natural philosophical-holistic interpretation of the meaning of life and a scientific-inductive interpretation of single biological processes are justifiable, though the limits of these two ways of consideration must always be respected.

II. Rhythm and Polarity – Physiological Analysis and Phenomenological Interpretation

> Associated things are unities: wholes and non-wholes, accord and discord, harmony and disharmony, all from one and one from all. Invisible harmony is more powerful than visible harmony.
>
> Heraclitus

> Forgetting is part of all activity: just as light and darkness are part of everything organic as of life.
> Nietzsche, Impromptu Reflections
> (Unzeitgemäße Betrachtungen)

Biology must be concerned with rhythm and polarity since they are fundamental features of life (cf. p.64f.). There are two ways to gain knowledge and understanding of these basic peculiarities. On the one hand, rhythm and polarity can be interpreted as living phenomena in order to perceive their essence and meaning. On the other hand, one can attempt to comprehend periodically varying living processes in causal terms, i.e. to explore and clarify their causes, the conditions in which they arise and the rules which they obey. In the first case, a figurative, qualitative mode of representation is chosen, while functional, quantitative criteria play the predominant role in the second case.

Contrasting the two possibilities of understanding emphasizes the wide perspectives in which this problem of living research must be seen. One should attempt to illustrate the two ways of thought with a few examples in order to show what findings and conclusions they can produce.

We shall first treat the question as to how the rhythm
and polarity of sleeping and waking (common experiences
of man) can be interpreted on the one hand according to
the conditions in which they may arise and according to
their essence and meaning on the other. It will be clear
immediately that the question as to the *reason* for and
conditions in which a rhythmical process may arise, i.e.
the causal question, is fundamentally different from the
problem of the *meaning* of this phenomenon.

We shall first traverse the first path and see what results can be found. Afterwards we shall choose the second
path in order to be able to compare critically the knowledge gained with the results of the first method.

1. Causal Analysis of Periodic Processes

1.1. Phases of Rest and Activity in Animals

If the bodily movements of animals maintained in an artificial light-darkness cycle are registered quantitatively
for a series of days (*Aschoff* et al. have carried out
such studies on chaffinches), it is noticed that the birds
abruptly wake up when the light in their cages is turned
on and develop a lively activity which can be continuously
measured by the number of hopping movements (*Fig. 2*). After an initially high level, activity decreases in the
course of the artificial 12-hour day and passes into a
phase of rest after the light is switched off. The rest
phase is occasionally interrupted by a few movements at
the beginning, but then passes in almost complete rest.
The activity curves are almost the same on different days,
though every day there is somewhat different development
and also a certain variability in the time at which activity begins.

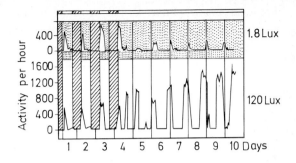

Fig. 2. Periodicity of the activity of a chaffinch in artificially alternated light and darkness and in continuous light of 1.8 and 120 lux. Lightly shaded phase = artificial day with light of 1.8 lux; darkly shaded = artificial night; unshaded = artificial daylight of 120 lux. Beginning of the two curves = light-darkness change; continuation = continuous light (after *Aschoff* et al., 1963)

If the animals are now exposed to continuous light instead of the artificial alternation of day and night, then there is a highly interesting change in the activity curves. On the first days of continuous light, the activity was sharply reduced to a low level. However, the periodic activity then returned; the intervals were the same as in the initial phase although the periodicity could not have been induced by an alternation of light and darkness.

The intensity of the continuous light was important for the result; it affected the amount as well as the beginning and duration of the activity. When weak continuous light was used (about 1.8 lux), then the total activity, i.e. the integrated number of bodily movements per activity period, decreased although the beginning and duration of the periods of activity were the same as in the controls. Contrasting behavior was found with an intense continuous illumination with about 120 lux; the total activity increased from day to day compared to the dark-light regime. The periodicity was changed inasmuch as each activity phase began earlier than the last, i.e. the periodicity became increasingly frequent. The periods of activity also became much longer, so the resting time became shorter and shorter and finally the animals did not rest at all.

Aschoff was able to summarize the overall results of all his studies as follows (see also Fig. 3):
a) With increasing intensity of the continuous illumination, the frequency of the sleep-waking periodicity increased, so that the artificial periods ran in front of the normal day-night change (*Fig. 3*, upper graph).
b) The proportion of sleeping to waking time was increased in favor of the waking time with increasing intensity of illumination, i.e. the period of rest became shorter and shorter and was eventually abolished completely (*Fig. 3*, middle graph).
c) The amount of activity, i.e. the number of body movements integrated over the waking period, became increasingly larger with increase in intensity of illumination (*Fig. 3*, lower graph).

These results suggest that light plays a crucial role in inducing, stimulating and maintaining the activity of all living creatures. However, this is not so. There are animals whose activity phase begins in the night and which sleep during day. Still others live completely in darkness.

While *Aschoff* found the behavior which has been described was continued in a series of diurnally active animals, there was an entirely different picture in nocturnal animals. If he chose a dark-active animal, e.g. a mouse or a golden hamster, the following behavior could be observed (*Fig. 4*): the periods became shorter with the decreasing intensity of continuous light and both the amount and duration of activity were increased. It could consequently be concluded that the illumination is not the only decisive "factor" in forming the vital rhythm of sleeping-waking behavior of living organisms.

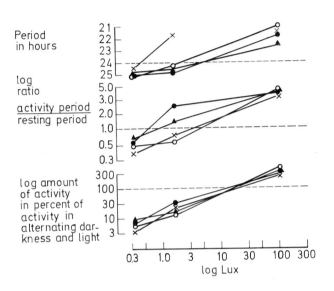

Fig. 3. Parameters of the periodicity of activity of chaffinches as affected by the intensity of illumination (after *Aschoff* et al., 1963)

In order to give a reason for the behavior of the animals, *Aschoff* and his collaborators tried to interpret the transition from the waking state to a period of sleep as an "endogenous oscillation". They assumed that the periodicity is based on a "continuous process" which is divided into two parts by a threshold.

When the rising periodic function exceeds the threshold, activity is initiated. However, if the function falls

below the threshold again after about half a period, the
activity phase ends. The threshold divides the quantatively different functional values into two qualitatively different states, the activity phase and the rest phase [1].
The position of the threshold in relation to the average
value of the oscillation (also called the "zero value"
or "level") determines both the amount of activity and
the proportion of activity to rest [2]. If the threshold is
low, then the physiological state entails raised excitability and the amount of activity is greater and the active period longer than the rest period.

If it is assumed that the incident light affects both the
threshold and the frequency of the periodicity, then the
situation with respect to the weak and intense continuous
illumination can be described roughly as in *Figure 5*. In
diurnally active animals, a positive influence on frequency and threshold (= "excitability") is to be assumed,
i.e. the frequency is raised and the threshold lowered.
In nocturnally active animals, on the other hand, there
is a negative effect, i.e. the frequency is lowered and
the threshold raised. *Wever*, one of *Aschoff*'s collaborators, postulated that the parameters frequency and threshold are mutually linked. *Aschoff* further discusses the
"biological clock" controlling the periodicity of activity and rest. He assumes that both the daily alternation
of light and darkness for the animals studied and also
environmental temperature fluctuations (if of adequate
intensity) are important factors.

---

[1] In such a causal analysis, one is clearly forced to describe the "polarity" of sleeping and waking with a "continuous", quantitative function.

[2] It appears more appropriate to take the "zero value" as the relatively constant value and to allow the "threshold" to vary as the reciprocal of excitability. The nomenclature was accordingly changed.

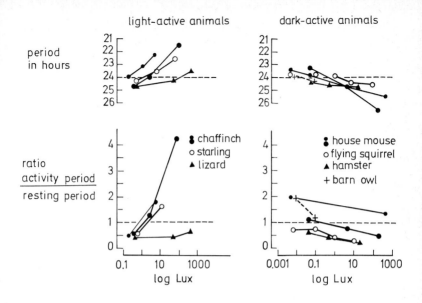

Fig. 4. Periods and proportions of activity to rest as influenced by the intensity of illumination for diurnally and nocturnally active animals (after *Aschoff* et al., 1963)

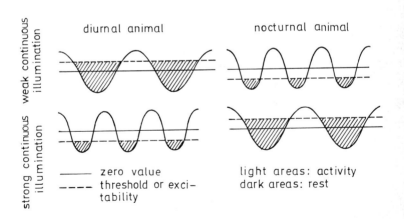

Fig. 5. Spontaneous periodicity of activity in continuous light: diurnally and nocturnally active animals (modified after *Aschoff*)

1.2. Phases of Activity and Rest in Man

Finally, to determine the *waking* and *resting phases in man* under accurately controlled conditions, *Aschoff* et al. also carried out investigations in a soundproof, windowless subterranean living and sleeping space. The test subjects, who did not have a clock, were allowed to determine when they turned off the light and went to sleep. They also decided when they turned on the light after waking up and how to arrange their day (including preparation of meals). Such studies showed (*Fig. 6*) that the behavior of the test persons remained periodic to a certain extent even if they did not have an external "Zeitgeber". The time spent awake remained almost the same although the time of waking up was delayed more and more with each day. If the waking times of the test subjects are shown below each other, the time of waking up after 18 days is shifted by 32.5 hours compared to its position at the start of the experiment.

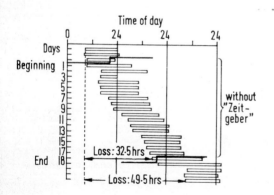

Fig. 6. Waking times of a test subject after exclusion of all "Zeitgeber" (after *Aschoff* and *Wever*, 1962)

After leaving the test room, synchronization with local time was reattained by a further slowing of the periodicity over two extremely long waking periods. The duration of the periods of all test persons was between 24.7 and 26.0 hours, a finding which the authors attributed to the relatively low intensity of illumination [3]. The observation is worth mentioning that the first days without time reference were generally found to be unpleasant and that the test persons had reported a feeling of excellent wellbeing when the regularly delayed periodicity set in. The initial interest in the "real" time is completely lost.

Quite similar observations can be made in periodically active organs, e.g., the heart, as for the periodicity of the activity or waking phases and the resting phase correlated with these in animals and humans.

The *activity and recovery phases of the pulsating heart*, the phases of spreading excitation and subsidence of excitation of the heart muscle can be measured in low temperature studies on dogs (*Blasius* et al.).

The animals are lightly anesthetized with hexabarbitone-suxamethonium with artificial respiration and are introduced into a low-temperature chamber, maintained at an average air temperature of $-12°$ C. Under these conditions the body temperature of the dogs regularly fell from $37°$ C to $22°$ C. If electrocardiograms are similarly taken from the chest and from the extremities, the duration of the spreading of excitation in the heart muscle (duration of QRS time) increased by about 50% with the falling deep-body temperature of the animal (*Fig. 7*). The phase

---

[3] Of course, the authors discuss also other possibilities (*Aschoff* and *Wever*): Recently *Aschoff* observed that any environmental stimulus can be a "Zeitgeber" (1966).

of returning excitability, the repolarization phase (or ST time) was very much more attenuated (about 265%) if the deep-body temperature fell to 22° C. However, the heart period was delayed the most markedly by cooling (about 470%) [4].

The amount of cardiac activity remained almost the same during the fall of temperature since the absolute spatial vector of QRS did not increase significantly [5]. However, the relationship of the duration of the depolarization phase (QRS duration) to the repolarization phase (ST duration) is substantially changed in favor of the latter.

It is also possible to view the periodic alternation of depolarization and repolarization as an oscillation process (*Fig. 8*). In the same way, it must be assumed that both excitability, i.e. the depolarization threshold, and frequency of oscillation depend on the temperature and are related to the zero value of the periodic function. The depolarization threshold divides the periodic function into an activity phase and a resting phase. Taking the time values for QRS found at 37° C and 22° C deep-body temperature, calculations based on a cosine function show that the threshold is substantially raised by lowering the temperature (*Blasius*); this would explain the long repolarization times. Moreover, it is clear that the che-

---

[4] The temperature coefficients (Q_{10} values) in the temperature range studied were 1.5 for QRS, 2.6 for ST and 3.2 for the heart period RR. Consequently, the processes during the spreading excitation phase (QRS period) are mainly physical and those during the recovery phase (ST period) chiefly chemical processes which are influenced by temperature (*Blasius* et al., 1957; *Blasius* et al., 1961). It is worth mentioning in this connection that raising the temperature regularly leads to shortening of the QRS, ST and RR times (*Ulmer* et al.).

[5] It appears justified in this context to characterize the absolute production of potential by the heart in the phase of spreading excitation as an "activity quantity".

mical processes during repolarization constitute the major part of the energetic processes in the heart muscle during the deep cooling phase, a result which can be reconciled with other physiological facts.

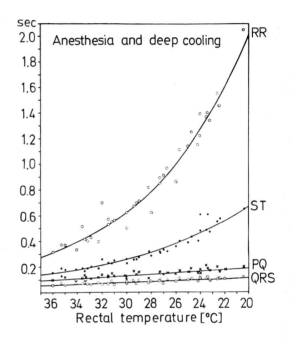

Fig. 7. Deep-cooling experiment in dogs: spread of excitation and recovery of heart muscle (*Blasius* et al., 1961)

2. A Critical Analysis of a Causal-Analytic Interpretation of Rhythmic Phenomena

After these causal analytic interpretations of periodic processes which utilize the physical model of an oscillation process, we shall now attempt critically to describe how far the causal analytic interpretations deviate crucially from living reality.

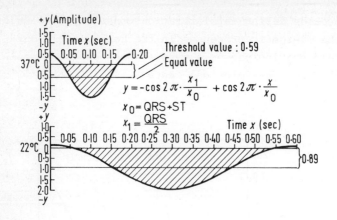

Fig. 8. Depolarization (light areas) and repolarization (dark areas) as an oscillation process in excitation of dog heart muscle. Above: at normal body temperature; below: in hypothermia (*Blasius* and *Repges*, 1968)

We do not intend to dispute or attack the experimental results of *Aschoff* which have been described. There is no doubt that the experiments were flawlessly and precisely carried out and evaluated according to the rules of physiological methodology. What is in question is whether the experimental conditions and the significance of the results are relevant for the assessment of living reality.

It is basically a question as to the value of scientific experimentation. Without doubt this is a controversial topic, but it must be treated openly and honestly; only in this way can epistemological clarity be attained.

It constitutes a substantial restriction of reality (which is always made and accepted without comment in an experimental treatment of living phenomena) to carry out studies on animals (in our case, chaffinches) in a cage. Free movement is restricted; in particular flying is made impossible and the whole natural environment is excluded. It is also a very abstract condition that the animals are isolated, that they are removed from the living interaction with their own species and sex partners and that they are also isolated from the rest of the environment. It is a further simplification of the living conditions to remove the animals from the changing conditions of natural weather with its fluctuating temperature, humidity, and light intensity. Neither the seasonal fluctuations in incident light and day length nor the slow transition from darkness to light and back were taken in account. Instead of this, every day full light for 12 hours and then absolute darkness for 12 hours were automatically turned on at constant intervals. The allocation of the food also remained constant for 6 days, so that no activity could be observed during food-seeking behavior.

The life of the birds in such cages is highly artificial (comparable to conditioned reflex behavior or to taming) and by no means corresponds to their natural life. Expressed simply: the soul of the animals which is only comprehensible through their interaction with their frequently changing environment is entirely left out in these investigations.

Quite apart from these crucial considerations there are a number of other events in the experimental setup which must be borne in mind and require special criticism.

If continuous light was turned on in the first "days" after the light-dark illumination, activity initially diminished to a quite low level. If (as the authors assume) incident light were the crucial "Zeitgeber" during these studies on chaffinches, such a phenomenon should not arise. Even with a low level of illumination, a difference between the level of initial activity and that after continuous illumination should not be detectable.

When continuous illumination of a higher intensity was turned on, the frequency of the periodic activity increased. If an "internal clock" ("innerer Zeitgeber") is postulated (for example, periodic starting of a metabolic process) it would not be clear why this periodicity should suddenly be disrupted by the influence of light. At the most, it could be stated that light also affected the recovery processes in this case; these were accelerated, increasing the periodicity. However, this causal explanation would require light to be a permanent influence, not to be an image effect (or trigger = "einmaliger Zeitgeber").

One can therefore conclude from these two effects that neither an "external" nor an "internal clock" which is completely independent can be postulated. The real rhythmic process is thus of an entirely different nature; it cannot be interpreted in physical terms.

A further difficulty arises if the increase in frequency of periodic activity and the reduction of the resting period with increasing amounts of activity, which are observed with continuous illumination of high intensity, are to be explained as physical processes. There is obviously a change in the performance of the animals here; periodicity is gradually replaced by continuous activity.

It can be assumed that the normal dynamic equilibrium of energy uptake and release is disturbed and that an irreversible change with increase of entropy of the energetic processes takes place. It can be conjectured that with constant continuous illumination the animals would finally die. This is not expressly reported by *Aschoff*. At the very least the conditions in these studies are quite abnormal.

It must be further taken into account that constant continuous illumination of an intensity such as was used in these studies is not found on the earth; therefore the chaffinches live in their cages under conditions which do not occur naturally.

Even if one assumes the illumination in strong sunlight to be very great, the birds would be able to recover from such a flood of light in the following night or be able to flee into the shade of a forest. This is to say that the studies were carried out under conditions which do not occur on earth at present and have never occurred in the past.

As one never will be able to find an "ultimate cause" or a "trigger" ("Zeitgeber") for the periodicity of living organisms by scientific experimentation, any explanation is rendered unacceptable when considered in the light of the completely different results with respect to the amount and duration of activity and to rest in diurnal animals on the one hand, in nocturnal animals on the other If one wished to refer the periodicity to a single "factor", i.e. a measure of energy, then this factor would have to be active at one time and inactive at another; but here the causal principle breaks down completely.

Interpretation of rhythmic phenomena in terms of oscillation processes can partially describe an isolated process. However, this description has nothing to do with a causal explanation. The natural events of movement and rest are qualitatively so diverse that they cannot be "explained" by an uniform, "continuous" process. The same applied for interpretations of the depolarization and repolarization process in the heart under conditions of hypothermia. Investigations of conditions such as anesthesia, paralysis of muscles by curare, artificial respiration and deep cooling have practical medical significance, but these conditions are irrelevant for the normal functioning of the organism.

Studies of human behavior and exclusion of natural lighting are also of practical value for critical examination of certain conditions of life and work. However, the experimental conditions are artificial and more attuned to a prisoner's existence than to the free vital interchange of a human being with the varying influences of his environment.

3. Phenomenological Description of Rhythm and Polarity

What is the essence of the rhythm? The foregoing arguments permit the important conclusion that rhythm is a phenomenon of life, but that it cannot be explained in terms of "energetic" causes. It is consequently not permissible to identify the vital rhythm with a physical oscillation process. We therefore cross the threshold into metaphysics where the rhythmic soul-body connections are to be treated separately from mental acts.

If we wish to describe the essence of the vital rhythm, then we can say that rhythm is the most general vital peculiarity shown by all living creatures, including man. If we view the whole universe as living and animate, as did the Greeks, then rhythm is indeed to be looked upon as a cosmic phenomenon.

As shown by *Ludwig Klages* in his fundamental investigations [6], regularity or tempo ("Takt") stands in contrast to rhythm. The former are seen as feats of human intellect and will.

Rhythm, said *Klages*, can appear in its complete form ("Gestalt") in entire absence of regularity or tempo. On the other hand, tempo (this is very important and significant) cannot appear without cooperative mediation of rhythm. The decisive difference between rhythm on the one hand and regularity or tempo on the other is that rhythm represents a "renewal of the similar" while regularity or tempo represents "a repetition of the same".

Rhythm is consequently the appearance of living organisms possessing a soul, while tempo is an expression of the spirit and the will, which is an exclusive prerogative of man and can only be produced by man (cf. *Table 3*).

Table 3. Rhythm and Period (according to *Klages*)

	Rhythm	Period or tact
Origin	General phenomenon of life	Expression of human spirit or will
Precondition	Polarity of body and soul	Dualism of spirit and life (life = body-soul polarity)
Essence	Renewal of the similar	Repetition of the same
Domain	Kingdom of appearing forms (Gestalten)	Region of spiritual acts

Rhythm is a polarized permanence as *Klages* further says; this means that rhythm as a phenomenon of life always represents a polar event. The actual essence of the vital

---

[6] In the recently published monograph by *A. Sollberger*: "Biological Rhythm Research" (Amsterdam-London-New York, 1965), the work of L. *Klages* "On the Essence of Rhythm", is mentioned among 3,100 references on the subject, but its essential and generally valid content is not discussed.

phenomenon is expressed in polarity. Polarity consists for example, between the movement of a soul and the reality moving this soul, or, in our case: between the rhythmic phenomenon of animated repose and of waking or movement replacing this repose (*Blasius*, 1968).

As argued earlier with reference to *Klages*, the soul of an animal and man has the ability to contemplate; that is, the ability to fuse the active images of the phenomenal world with the images present in the soul. The essence of the polarity of experiencing soul and active image comes alive in this fusion.

Polarity is further expressed by the fact that the connection between image and soul can be followed again by an alienation ("Entfremdung"). According to *Klages*, alienation and connection of figurative soul ("Bildseele") and phenomenal image are in equilibrium in the contemplative state.

The rhythm of sleeping and waking may also be viewed as analogous to these phenomena. In the waking state the contemplative capacity of the soul is chiefly concentrated on the surrounding reality. However, the internal images are contemplated during sleep. Fusion with external reality is most pronounced during the phase of greatest wakefulness, while alienation from this reality occurs during sleep. On the other hand, equilibrium of fusion and alienation is reached at rest during the waking state, when the contemplative capacity is consequently at its most pronounced.

This behavior can be observed in animals and man. We conclude that polarities constitute essential prerequisites for rhythmic phenomena of life and that these phenomena cannot be interpreted otherwise than in terms of polarities.

What is the effect of regularity or tempo on rhythm? Without doubt, timing can influence rhythm, strengthening it. *Klages* comments on this: "Despite the fundamentally different origins of rhythm and timing in man the two can fuse with one another: even an existing rhythm can be intensified in its action by tempo".

In the chaffinch studies which were described the intensified periodicity under more intense illumination could be interpreted as an analogous phenomenon. Here the periodicity was intensified by an environmental influence such as was indeed artificially produced by a human agency.

Since every causal analytical interpretation of the phenomena of life aims at establishing rules and principles, it must inevitably disregard entirely the "living renewal of the similar" (i.e. rhythm and the soul), and attempt to bring the living creature into a state in which a "repetition of the same" is possible, at least under certain conditions. The empirical scientist is usually satisfied with this result because it represents the end of his analysis. If he wishes to postulate a "factor", a "cause" or a "Zeitgeber" and trace this back to other causes, the experimental conditions must be modified. He falls into a "regressus in infinitum". An "ultimate cause" for a living process cannot be found.

Consequently, an empirical-scientifically orientated analysis of a vital phenomenon, such as rhythm, can only yield a result reflecting the rules of the methodology which has been applied, but which is entirely independent of living reality.

4. Summary

A critical analysis of the results of physiological experiments which demonstrated periodic processes clearly shows that, basically, causal analysis can only yield rules and laws resulting from specific experimental conditions. On the other hand, in a completely natural environment living creatures are a rhythmic and polar part of the changing phenomena of their environment. All activities such as their behavior in sleep and waking, in rest and in movement, appear as a rhythmic interaction with their environment in the widest sense. As the "renewal of the similar", rhythmic life is to be contrasted with tempo or periodicity, "repetition of the same". Life remains inexplicable and unquantifiable in physical terms: most fundamentally, it is to be viewed not as the effect of a cause, but as a mystery.

III. Bodily Movement and Exercise – Physiological Analysis and Phenomenological Interpretation

> His walking erect distinguishes man
> from the animals. Man's learning
> to hold the center of gravity of his
> body high above a small pivot con-
> stituted the beginning of the release
> of his body from gravity. Dancing,
> which is common to all peoples, is
> a refinement of this capacity for
> artistic expression.
> Ludwig Klages

1. Phenomenological Interpretation of Bodily Movement

The Greek physician of antiquity, *Alkmaion*, made the pregnant statement about life, that it is αωτοκινητος, "autonomously motile". Even today, this important insight can serve as a principle for understanding the essence of movement, the exercise of movement and even the healing effect of movement.

If life is "spontaneous movement", then loss of the capacity to move is to be viewed as a loss of vitality. Every illness can then be interpreted as loss of movement. This would imply and at the same time necessitate that illnesses in general should be seen in terms of a loss of movement. This idea has up to now not yet been consistently pursued. It is not only a matter of loss of mobility of the actual organs of movement - the muscles, ligaments, joints, cartilage and bones - but also the defective movement of the internal organs - the heart, the vascular system, the lungs, the viscera, the glands and the nervous system. All the pathological changes which can be observed in an organism can in the final analysis be viewed as losses of movement. Logically, all curative measures should be viewed as measures to restore mobility and movement.

The question now arises as to how this living movement is to be described. The Greeks have also provided the correct and valid answer to this question: "all life is rhythmic movement" said the Greek natural philosopher *Heraclitus*. This means that life is not to be viewed, for example, as a continuous, steady movement but as a rhythmic, pulsating movement. Rhythm is flow, but not continuous flow; it alternatively rises and ebbs (cf. Chap. II, Table 3). This view is also extremely important in the therapeutic use of movement.

2. Physiological Analysis of Bodily Movements

After developing these holistic ("ganzheitlichen") principles, one should first attempt to analyze the individual elements of the organism which appear particularly important for the mobility and movements of the body. This analytical or natural scientific representation will give rise to certain information for particular circumstances which must be precisely characterized as experimental conditions. However, it must be emphasized that the information gained is always prejudiced by the methods used in the analysis, and is thus limited in its applicability. If this fact is made sufficiently clear, one will not make the mistake of generalizing or overrating the results gained by analysis or taking them for the essence of the vital phenomenon itself. This important distinction was discussed in more detail in Chapters 1, 2 and 3 (*Blasius*, 1962).

A few fundamental facts which show the possibilities of making scientific statements about the effects of bodily movement and exercise on the organism should be described here.

Among these are: the external movement of the organism with the interplay of muscles and nervous system, the changes in the organ systems of the body, the conditions of bodily exercise and adaptation, and the concomitant functional changes of the important systems of internal movement: circulation, respiration and metabolism. Finally, the therapeutic use of physical exercise will be discussed.

2.1. The External Movement of the Organism in the Interplay of Muscles and Nervous System

2.1.1. The Basic Properties of Muscles

In order to understand the interplay of muscles and nervous system, it is necessary to mention some basic properties of the individual elements, firstly muscle: Muscle is characterized by three properties. It possesses: 1. extensibility, 2. elasticity and 3. contractility.

1. The *extensibility* of isolated muscles can easily be demonstrated by loading them with increasing weights (length-tension curve, *Fig. 9*). The most important characteristic of the resting stretch curve is that the extensibility is not proportional to the load as is an elastic band, but decreases with increasing load; the elastic resistance increases. The muscle tissue breaks at a certain maximum load; this is the point of maximal muscular strength [size order of the breaking strength: 10 kg/cm^2 transverse section of the muscle tissue; variation from

animal to animal (frog: 3 kg/cm^2); differences between individual human beings: decrease with age; danger of overloading the muscle].

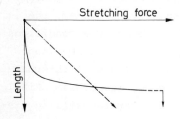

Fig. 9. Rest stretching curve of skeletal muscle (continuous curve; arrow show the point of maximum muscle strength = limit of breaking strength). Broken line = stretching of a rubber band. (*Blasius*, 1970)

2. *Elasticity*. After stretching, the muscle returns to its initial length as a result of its elasticity or its resetting capacity, although the return is not complete.

In an investigation of stretching and relaxation, curves were found (*Fig. 10*) which do not coincide: the relaxation curve is always shifted in favor of greater length or lower tension. The area between the two curves, the so-called hysteresis area, corresponds to the loss of work during the stretching and relaxation cycle. The hysteresis area decreases in the course of several stretching cycles. This indicates that the work loss decreases with increase in the number of stretchings i.e., stretching and relaxation proceed more economically.

3. *Contractility*. Tendons only show extensibility and elasticity, while muscle also has the capacity to contract. With increasing load, every muscle can contract

either isotonically, isometrically or, as in most cases, in a mixture of the two forms of contraction, i.e. auxotonically.

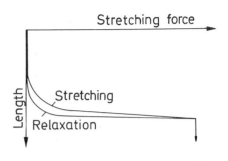

Fig. 10. Stretching and relaxation curve of a resting muscle (*Blasius*, 1970)

If one wishes to compare the muscle performance in the different forms of contraction, one starts with the resting extension curve (*Fig. 11*). If the isolated muscle is brought from a state of increasing stretch into an isotonic contraction, i.e. is permitted to undergo a change in length, then shortenings are produced which together constitute the *curve of isotonic maxima*.

If one arranges the experiment in such a way that the muscle can after a stretch change only its *tension* and not it length, then changes of tension are observed which together make up the *curve of isometric maxima*. In isometric contraction no external work is performed; the degree of effectiveness of contraction is zero. The isometric maxima are higher because no external work is performed. However, the energy which is released is consumed for internal work in the muscle (shifts within the filament system). In isotonic contraction the degree of effectiveness in unloaded (i.e. unstretched) muscle and at the point of absolute muscle strength is also zero (*Fig. 12*). In the area between the two points lie the shortening and work maxima. The shortening maximum is found with low loads, while maximum work is possible at moderate loading.

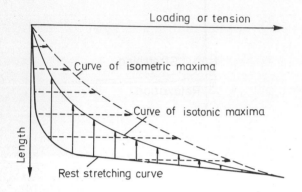

Fig. 11. Rest stretching curve of a muscle and curve of maximum muscle contraction under isotonic and isometric contraction conditions (*Blasius*, 1970)

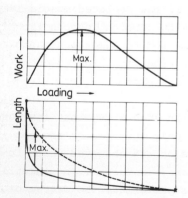

Fig. 12. Lower part: rest stretching curve (———) and curve of the isotonic maxima (---). Arrow = maximum shortening. Upper part: dependence of muscular work on loading in isotonic contraction (work = shortening x load). Arrow = maximum isotonic muscular activity (*Blasius*, 1970)

If the mobility of the muscles, the tendons and the joints is to be restored, then small loads are entirely adequate. If a good working capacity of the muscles is aimed for, not maximum but moderate loading is useful. However, high loading constitutes the precondition for the adaptation processes in the muscle (see below).

Analysis of muscle activity can therefore yield certain guidelines for exercise therapy. Of course these require a modification adapted to the specific muscle group in individual cases.

2.2. The Role of the Muscle Spindles in the Interplay of Muscles and Nervous System

Muscle contraction and relaxation are decisively influenced by the activity of special sense organs in the so-called muscle spindles which form the afferent connection with the central nervous system. The muscle spindles contain muscle fibers capable of contraction (*Fig. 13*); these are innervated by a special kind of thin motor nerve fibers (so-called gamma fibers). In the middle the spindle expands into a sack enclosing the nerve endings of a thick sensory nerve (so-called primary afferents).

Besides these primary afferent fibers, thinner sensory fibers, so-called secondary afferents, arise from the muscle spindle. The afferent fibers of the muscle spindle are stimulated when the main muscle is stretched. The excitation is conducted via the sensory nerve to the spinal cord and reaches the motor ganglion cells of the anterior horn. From there, the excitation passes via the motor nerves back to the muscle. This in turn is stimulated by the alpha motor endplates, i.e., caused to contract. A contraction arising from stretching the muscle, for example by a tap on its tendon or on the muscle itself, is a *monosynaptic jerk* ("*Eigenreflex*").

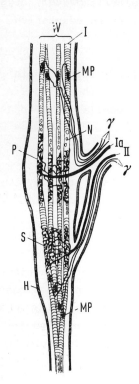

Fig. 13. Semischematic representation of a muscle spindle; the spindle is shown very much shortened. W = *Weismann* bundle, I = intrafusal muscle fiber, MP = motor endplate, N = sarcoplasmic nucleus, P = primary sensory ending, S = secondary sensory ending, Iα = afferent nerve fibers of the primary ending, II = afferent nerve fibers of the secondary ending, thin γ motor fibers, H = fibrous coat (after *Blasius*, 1962)

However, the muscle spindles do not only go into action to stabilize the body in sudden reactions: they also supply continuous series of impulses of a frequency related to the extension of the muscle; the frequency increases logarithmically with tension as caused by stretching. The particular frequency of afferent impulses thus corresponds to the state of mechanical stress in a muscle (which mainly depends on the angle at the joint) and also to a specific muscle activity, which is designated "*contractile or reflex muscle tone*". This muscle tone is greatest in the flexor muscles if the joint to be flexed is straight and in the extensor muscles if the joint is bent. Equilibrium between the tone of the two muscles, the flexors and the extensors, arises if the forces resulting from the states of tension of the two muscles are in balance. This situation can arise very easily, for example in a passive posture under water. Such a tonic equilibrium is further promoted by the abolition of the effect of gravity in water. If the water is warm, an initial state of relaxation will eventually reach an optimum, which is particularly favorable e.g. for underwater treatment.

Furthermore, it is important for the function of the muscle spindle that these organs are arranged parallel to the main muscle fibers (*Fig. 14*). This parallel arrangement results in removal of the load from the actual muscle spindles with every contraction of a muscle; their sensory activity is thereby reduced.

The muscular elements of the muscle spindles, which can be brought into action by the gamma fibers, exercise an important influence on the *regulation of the tension of the whole muscle*. By altering the tension of the muscle spindles, the gamma system is able to intervene continuously in the overall tension of the muscle by influencing the frequency of the impulses sent out from the spindle afferent. Any stronger excitation of the gamma fibers intensifies the initial stress to some extent: this affects the impulse frequency in the sensory fibers which in turn reflexively raise the overall muscle tension.

This mechanism enables the organism to modify muscle tone continuously under the influence of external factors. Sensory stimuli from the end organs of the skin or from other receptors as well as impulses from the higher centers of the central nervous system can reach the gamma motoneurons via interneurons in the spinal cord or by direct action on the motor ganglion cells of the ventral horn. In turn, the motoneurons activate the reflex arc described, changing the tension in the muscles. It is therefore possible for sensory stimuli from the outer world to influence in this way the tone of all the muscles: this influence is extremely important for mobility and muscle contraction.

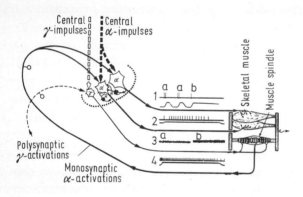

Fig. 14. Schema of the system of the α- and γ-motoneurons, their links with the muscles and muscle spindles and their central monosynaptic and polysynaptic reflex impulses. The curves signify: 1 = phasic α-stretch reflex in short (a) and longer (b) muscle stretching; 2 = tonic α-stretch reflex with sustained muscle stretching; 3 = spontaneous activity of a γ-fiber before (a) and after (b) central activation; 4 = discharge of muscle spindle with sustained muscle stretching (after *Blasius*, 1962)

As can be demonstrated experimentally, *all stimuli to the skin over a muscle* intensify the tone of the muscle, while at the same time these stimuli reduce the tone of other muscles. The tonic effects of a gentle stroking massage are thus explained, if one is clear about the conditions under which they occur most easily.

If the skin is cooled, there is in a triphasic occurrence an initial increase in tension of the muscles. There follows a partial relaxation in the adaptation phase and, if cooling is decreased, the wave-like increasing and decreasing tension phase (cold trembling) can be observed. Other skin stimuli, for example chemical stimuli, have a similar action; this is utilized in rubbing massages, poultices and baths (*Göpfert*)

Besides stimuli to the skin, the impulses received by the gamma motoneurons from the higher centers and from the cortex are very important for the activity of the whole muscle system. Via the muscle spindles, these central gamma stimuli bring about contractions which result in muscle movement: to a certain extent these derive from the contractile tone. It is significant that the contractions retain their fine graduation. In view of the knowledge gained about these physiological relationships, it must be considered certain that most voluntary movements proceed under regulation from the gamma system. A direct central excitation of the large alpha motoneurons alone is likely to occur in very rapid and violent movements.

2.3. Physiological and Morphological Changes in the Organism during Physical Exercise

In order truly to understand the effect of a physical excercise, it is necessary to know what physiological and morphological changes are taking place in the organism. I shall hence review the most important observations made during bodily exercise and the adaptation of the body to physical exercise.

"Every beginning is difficult" is a well-known saying. This particularly applies to starting physical exercise.

However, one observes a rapid rise in skill and stamina if the same physical exercise is repeated. This illuminates the difference between bodily exercise and the use of a machine. The body becomes more and more efficient through exercise and use while the efficiency of a machine becomes less and less. The initially low level of efficiency of physical work increases with exercise, while that of the machine decreases more and more.

This increase in physical performance arises from two quite different processes, *practice* and *adaptation*. Practice corresponds to the acquisition of a purposive interplay of the individual bodily organs, particularly the sense organs, the nervous system, and the muscles. Adaptation consists of a number of morphologically demonstrable changes in the exercised organs when work is repeated. The term "training" in sport is applied indiscriminately to the two processes.

In the acquisition of a bodily movement, the interplay between sense organs, e.g. the muscle spindles, the central nervous system, and the target organs must be practised. Every movement is carried out under the control of the sense organs. However, e.g. in walking this control by the eyes, the organs of touch and the muscle spindles etc. is totally unconscious because this movement is already practised. In a movement carried out for the first time every sensory impression must be consciously perceived in order to transform such impressions into a muscle excitation via the motor cortex.

With progressive practice, ever more unconsciously active sections of the central nervous system take over the overall control. For example, a child acquires the ability to write letters with conscious control of the individual hand movements, while the practiced writer can devote his attention chiefly to the content of what he is writing. In the same way, every exercise with a new movement content is initially possible only with conscious control of the target organs; the initial excessive efforts are more and more replaced by a more economic execution.

Practice is the necessary precondition for adaptation. The frequent use of central nervous pathways results in a channeling, while use of the musculature and the concomitant extra demands on the respiratory and circulatory systems constitute a strong stimulus for these organs, which is answered with appropriate processes of adaptation. On the one hand, the adaptation consists of a measurable growth of the organs. On the other hand, it consists of a change in their function. Together these bring about a corresponding increase in physical performance.

2.4. Muscular Adaptation

As shown by experiments on humans and animals, the musculature only grows with high performance, and not with long-lasting low-level performances. The transverse section of the muscle fibers increases, raising the efficiency of the whole muscle. There is also a proliferation of cell nuclei to which the lower fatiguability of the muscle is attributed.

Relatively short periods of daily exercise are sufficient to elicit growth if the musculature is fatigued only to the performance limit.

The Swedish physiologist *Johannson* proved that even a physically untrained person can greatly increase his performance. This classic study can be demonstrated by *Jo-*

hannson's own data. The study shows very convincingly what an increase in human efficiency is possible under physiological conditions.

Johannson lifted to the point of exhaustion 25 kilograms with both arms on an ergograph which permitted the height raised and the number of lifts to be recorded. Then he rested for three minutes and repeated the exercise twenty times in succession. The result of his efforts is shown in *Figure 15*. The initial performance was 4000 mkg/day. In the next two days the performance decreased almost to half; this was because of "sore muscles". These sore muscles were caused by a swelling and shortening of the muscle fibers which stimulated pain receptors. With further exercise and maintenance of a daily exercise program the performance rose in about 7 weeks to about 7 times the initial value. This is an astonishing increase in performance! These experiments have been repeated by other authors with the same result.

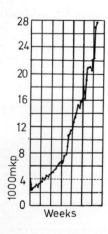

Fig. 15. Increase in performance in physical exercise (after *Johannson*, see *Blasius*, 1970)

A further interesting observation made by *Johannson* was that 50% of the adaptation gain was retained if the subjects exercised only once a week, and up to 30% if the subject exercised only every fortnight. Indeed, this training effect was retained if the subject exercised only once a month (*Fig. 16*).

There was a rapid disappearance of the muscle adaptation if the muscles were not used, and adaptation slowly disappeared completely.

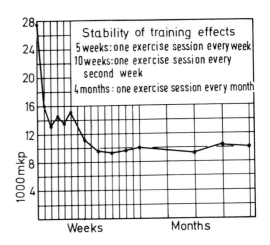

Fig. 16. Stability of training effect in the same exercises as those shown in Fig. 15 (after *Johannson*, see *Blasius*, 1970)

2.5. Adaptation of Heart Muscle

The heart and all vascular muscles also show a marked growth of the muscle fibers and thus an increase in weight with repeated physical work. This increase in weight is determined by the particular life style. Analogous increases in the weight of the heart muscle are also found in the animal kingdom. This fact is made clear by a table in which the relative heart weights of closely related animals with very different ways of life are given as a percentage of the body weight (*Ranke, Table 4*).

Table 4. Heart weights of some animal species as percentage of their body weight with different physical demands (after *Ranke*)

Domestic rabbit	2.40
Wild rabbit	2.76
Hare	7.75
Domestic duck	6.98
Wild duck	11.02
Moorland white grouse	11.08
Alpine white grouse	16.30

Extensive studies have shown that the number of the muscle fibers has not changed with increase of the ventricular volume of the human heart under conditions of physical exercise; rather, the fibers increase regularly in length and thickness in a "harmonic hypertrophy" (*Gauer, Fig. 17*).

A particular stimulus to heart growth is known to be provided by kinds of exercise which are associated with highest performances maintained for a long time; enlarged hearts are hence found amongst sportsmen, particularly in racing cyclists, cross-country skiers, marathon runners and professional rowers. However, such increases in heart size regress when the exercise ceases, provided that the limit of "harmonic hypertrophy" has not been exceeded.

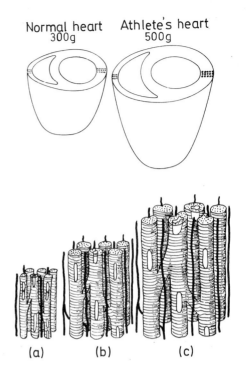

Fig. 17. Schematic representation (approximately natural scale) of the growth of the heart muscle fibers. a) infant heart. b) normal heart. c) athlete's heart. In the transition from b to c the number of muscle fibers does not change. The individual fibers only become longer and thicker (harmonic growth). The coarser tissue causes a thicker ventricular wall and a larger cavity (above) Note that each muscle fiber has its own capillary. In early infancy the muscle fibers are so thin that two fibers can be supplied from one capillary (after *Gauer*, 1960)

2.6. Adaptation of the Blood

A further adaptive measure of the body exercise is an increase in the red blood cells, which constitutes an improvement of the oxygen-carrying capacity. In particular, however, the capacity to bind carbon dioxide in the blood increases, i.e. the hemoglobin, the blood proteins, and the bicarbonate increase. Together these raise the capacity of the body to bind larger quantities of carbon dioxide.

2.7. Functional Adaptation

2.7.1. Circulation

The morphological changes are accompanied by adaptive functional changes in the internal organs. These will be treated briefly here.

Even the resting pulse is known to be markedly lower in physically trained than in untrained persons. While an average resting pulse of 72 beats per minute is found in adults, a resting pulse between 50 and 60 beats per minute is observed in physically trained sportsmen and heavy manual workers.

The difference is just as clear in an exercise study: the untrained subject raises his heart minute volume mainly by increasing the number of beats while the trained person increases the stroke volume with a correspondingly smaller elevation of the heart rate. The increase in the stroke volume is derived from the enlarged systolic reserve volume (*Fig. 18*).

Finally, after exertion the pulse settles down considerably more quickly in trained persons than in untrained persons.

2.7.2. Respiration

Breathing at rest is also slowed in trained subjects. The respiratory minute volumes decrease with adaptation in subjects used to work, so that the trained person attains the same oxygen uptake with a smaller respiratory expenditure than the untrained subject (*Ranke*). These relationships are illustrated in the next figure (*Fig. 19*).

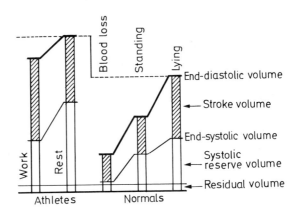

Fig. 18. Stroke volume and systolic reserve volume of physically trained persons at rest and after work and, for comparison, in normal subjects (lying, standing and after blood loss) (after *Gauer*, 1960)

2.7.3. Utilization

The trained and adapted person is accordingly able to make do with a smaller heart minute volume than the untrained person due to an enhanced utilization of the blood. This principle is interpreted in the following formula:

$$V_m \cdot \text{Utilization} = \text{Oxygen Consumption}$$

The relationships between heart minute volume V_m, arteriovenous oxygen difference (utilization) and total oxygen consumption of the body and increasing exercise are shown in *Figure 20*.

Even at rest, very low heart minute volumes are found in trained sportsmen (2,5 - 3,0 l/min) compared to untrained subjects (5,0 l/min) and correspondingly higher utilization values.

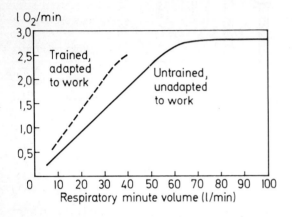

Fig. 19. O_2 uptake in relation to respiratory minute volume under conditions of exercise in untrained persons (———) and trained persons (----) (after *Ranke*, 1941)

2.7.4. Autonomic Nervous System

The adaptation of all these functions of circulation, respiration, and metabolism shows clearly that adaptation to physical work is accompanied by an adaptation in the autonomic nervous system. In this overall adaptation, the capacity of the organism to attain a state of higher order is expressed. The energetic side of this higher order is expressed in the improvement in the degree of working efficiency under conditions of physical exercise.

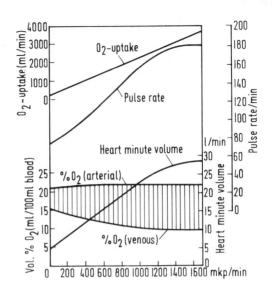

Fig. 20. Oxygen uptake, pulse rate, arterial and venous O_2-saturation and heart minute volume with increasing exertion on the ergometer (assembled from values in the literature, *Blasius*, 1970)

3. Exercise as a Therapeutic Measure

These arguments can be rationally supplemented by a few aspects relating to therapy with physical exercise.

On the one hand, such measures consist of passive exercises: massages, baths, underwater therapy and other physiotherapeutic effects. On the other hand, there are possibilities for active exercise: gymnastic exercises, brush massages, or walking and running.

The "obstacle" or "terrain" treatments are successfully applied in disorders of heart and circulation accompanied by reduced performance. They consist of movement exercises, in increasingly extensive walks and in climbing slopes of increasing gradient. Terrain cures are supplemented by physiotherapeutic exercises and for some time now also by hydrotherapy and psychotherapy. Modern terrain cures are characterized by relocation of treatment to the outdoors, maximum possible activity on the part of the patient group or individual therapy, measures designed for physical adaptation and thus improved efficiency.

The development of exercise treatment and, in particular, prophylaxis is highly illuminating. It is also reasonable in physiological terms. It confirms the opinion expressed at the beginning that illness is to be viewed quite generally as loss of movement and consequently, in the final analysis, that therapy aims at restoration of the mobility, movement and adaptation of the physical organs.

It is a highly rational therapeutic measure that rhythmic movement of all the bodily organs is aimed for in exercise treatment and that this rhythmicity is stimulated and increased by that of the environment: the countryside, the weather, light, water. Such treatment is highly successful in environments characterized by scenic beauty.

4. Summary

Proceeding from the fundamental principle of the Greek physician *Alkmaion* that "life derives the movement-giving impulse from itself", a loss of life, i.e. disease, can be defined in general terms as constituting a loss of movement. This not only concerns the loss of movement in the organs of motion, but also includes all the organs of the body. Thus, therapeutic measures aim at the restoration of mobility and movement. In this context, life must be understood as constituting a rhythmic and not a uniform movement.

Aside from a holistic interpretation of life, partial analyses of individual movement are also given. Their limited value is nevertheless emphasized.

The external movements of the organism in the interplay involving the sensory organs, the central nervous system, and the muscles are described. The basic characteristics of muscle, i.e. extensibility, elasticity, and contractility, are analyzed. Stretch curves, isometric and isotonic maxima are elucidated and the consequences of these natural laws are considered in relation to bodily movement and to exercise.

A special chapter is devoted to the role of the muscle spindles in controlling muscular movements. The importance of muscle spindles in relation to muscle tone, equilibrium in the tone of antagonistic muscles and the regulation of muscle tension is dealt with exhaustively and the possible significance for both active and passive bodily exercise is considered.

In addition, the morphological and physiological changes occurring in the organism under conditions of physical stress are enumerated. Here it is necessary to distinguish between those changes which occur at the training or exercise stage and those which take place during the adjustment stage. The processes of adjustment occurring in the muscles, myocardium and the blood as a result of physical exertion are outlined as are the principal changes in the circulation, breathing, metabolism, and the vegetative nervous system.

In the final section, physical exercise as a form of medical treatment is discussed. The importance of rhythmic movement of the body and the importance of environment are both particularly emphasized.

IV. Human Language – Physiological Analysis and Phenomenological Interpretation

> A word never has exactly the same meaning in two different sentences.
>
> Language retains its soul only in poetic usage.
>
> Ludwig Klages

Language is the most important essential human characteristic, distinguishing man from all other living creatures. This statement is based not only on the performance of profound human speakers, as for example the poet reading from his works, a great teacher who develops his experiences and thoughts before his audience, an actor who enthralls by words and their expressions, a speaker calling for action, or a loved one whose feelings impinge on our ear. We also come to this insight through experience with fellow human beings whose speech is disturbed, who have lost the power of speech, or have never possessed it from birth.

Language is such a basic human attribute that without it we would be beings without history like animals or plants. We would be unable to say anything about our experiences, thoughts, and actions. We would be unable to communicate these experiences to other human beings, to appreciate, or use or further develop essential cultural values. It is hence not surprising that man at an early stage began thinking about his special faculty of speech.

Poetic sayings and scientific explanations in the most diverse disciplines provide persuasive evidence of the ideas which men of all ages have had about language.

From the great profusion of such evidence it is not my task to select even the most important or to give a historic survey of all the sciences which are concerned with language. Rather, I am concerned to show the principal possibilities of describing human speech and its limitations.

Language has often been made the object of experimental research by physiologists, neurologists, and other natural scientists: all these studies have usually been hampered by a particular difficulty.

This difficulty, about which the investigators were often insufficiently clear, is based on the fact that human language is a being possessing a body, a soul, and a spirit. In his major philosophical work and in his last monograph "Die Sprache als Quell der Seelenkunde" ("Language as Source of Psychology", 2nd edition, 1959), *L. Klages* has dealt with this very clearly and thoroughly. According to *Klages*, any purely natural scientific interpretation of language can only deal with the body of language, i.e., with speech sounds and their origin. However, the mind and spirit of language, i.e., the meaning of speech and the concept of speech must remain outside the scope of natural scientific treatment. This observation, which will be justified in detail, can explain the particular obscurity of the terms: "speech center", "association center", "localization of linguistic understanding, of cognition and memory", which are used in all books on anatomy, physiology, and neurology. These terms were coined in ignorance of the real facts and consequently gave rise to confusion.

The fundamental systematic clarification brought by the philosophical and in particular the phenomenological theories of this field is now slowly being recognized, even by natural scientists.

For the elaboration of the epistemological foundations of sciences such as biology, physiology, neurology and experimental psychology, which are today chiefly orientated in a natural scientific direction, and for the philosophy to the extent that they are concerned with phenomenological questions, it is hence highly useful to discuss thoroughly the problem of language from both a physiological and a phenomenological viewpoint in the light of *Klages*' theories.

Firstly, the results of natural scientific analysis of human language and its physiological basis should be presented. The areas in which phenomenological research can be applied should then be shown. Contrasting the results and validity of the two approaches will make it clear that such a study of language leads to insights which can also substantially facilitate access to special problems.

As argued in Chapter 1 and elsewhere, study of the phenomena of life - and language is a phenomenon of life - can make use of two fundamentally different approaches: on the one hand, the natural philosophical or figurative-intuitive approach and on the other hand, the natural scientific or experimental-analytical approach. The first can be designated "phenomenology" and the latter "causal theory".

If natural scientific research into life merely used the methods of natural science, then physiological investigation of language would only be concerned with the causal association between the physical processes of hearing and speech. It will become clear during the discussion of these processes that there is repeated deviation from this methodological condition.

1. Physiological Analysis of the Sound of Speech and of Hearing

1.1. The Activity Cycle of Hearing and Speech

If an organism interacts as a whole with its environment, then this always occurs through a specific activity of one or several of its many muscles. In the same way human speech is linked to the activity of certain muscles in the larynx and others which act on the rib cage in order to provide the larynx with the necessary air. However, the activity of the organs of speech would not be possible and also not have any meaning if the speech process was not preceded by hearing of words and sentences apprehended in terms of their meaning. The living organism has learned these words as images and they remain available to him.

It is already clear that the relationships are not simple. Consequently it was a long time before the role of auditory stimulation in hearing was known. The excitation proceeds from the ear along the nerve pathways to the cerebral cortex, where the auditory impressions are perceived. Speech impulses also emanate from the cerebral cortex; they are conducted via the nerves to the muscles in order to set these in movement, giving rise to speech sounds. These regions of the cerebral cortex which are responsible for auditory reception and which develop the impulses for speech are named "sensory and motor projection fields". Nerve pathways connect these fields.

The overall course of the physical processes involved in hearing and speech can be represented as an "activity cycle". This is set in motion by particular environmental stimuli (*Fig. 21*).

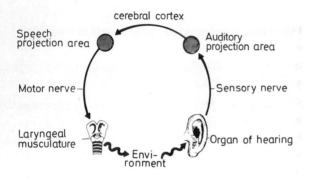

Fig. 21. Activity cycle of hearing and speaking, released by a sound from the environment and answered by a vocal sound (*Blasius*, 1967)

When one person calls something to another who hears and answers the call (either immediately or after a delay), then the excitation process in the answering person takes the course shown in the sketch. Nervous impulses pass from a sense organ, the ear, via the sensory and motor projection fields to the executive organs, the larynx and the muscles of the rib cage. This process can be represented as a purely physical process with physical methods, e.g. by demonstration of the excitation which is expressed in action potentials.

In order to consider the individual processes of this activity cycle in more detail, we begin with the *speech process*. If we register the excitation of the laryngeal musculature with the aid of the action potentials and simultaneously record the sounds according to oscillation frequency and intensity, then we obtain the following diagram (*Katsuki, Fig. 22*). Certain fluctuations of potential in the muscle can be detected at rest; these are indicative of a *resting tonus* (cf. Chap. III, 2.2). If the laryngeal muscles become active there is a sudden increase of frequency of discharge leading to a higher average level of recordable muscle potential. However, the audible sound only begins a short time later, which is evidence that the impulses had already arrived from the cerebral cortex before the muscles became active. This process is termed "inner speech": it could perhaps better be termed "speech-preparedness of the muscles".

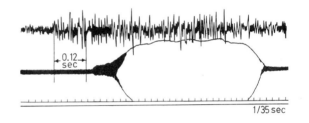

Fig. 22. Resting tone and action potentials from a laryngeal muscle (musculus cricothyreoideus) before and during singing of a high note (adult man). Above: action potentials (100 - 120 Hz). Middle: vocal sound, note ais[1] (461 Hz) (amplitude of oscillation reproduced in outline). Notice that the action potentials (designated "inner singing") begin 0,12 sec before the start of the sound vibrations (after *Katsuki*, 1950)

A point worth emphasizing is that during the process of speech formation the frequency of the action potential of the muscle, which is about 100 to 120 Hz, does not correspond at all with the frequency of the sound produced by singing (461 Hz). In production of speech, a transformation of the frequency of the stimulated muscle into that of the oscillating parts of the laranx, i.e. the vocal cords and the oscillating air spaces, is consequently necessary.

The *sound waves* emitted in *speaking* can also be registered and reproduced as notes, tunes and noises. Nowadays, the recording tape is a well-known aid for recording, storing and reproducing language. In one of *Grützmacher*'s earlier studies, the sequence of notes was measured with an oscillograph and a so-called melody-writer (*Fig. 23*). It is known that uttering the word "Leben" gives a characteristic temporal sequence of sound amplitudes. Of course, the "meaning" cannot be read off the curve. The words "Reben", "weben", "kleben" among others would give a similar sequence of sound amplitudes. A more precise analysis would certainly allow total differences to be perceived in the graph registered. However, what does not appear in the curve is the "meaning" which the speaker has given to the word "Leben": whether he had thought of "Leben" in a historical, physiological, economic, poetic, personal or in another sense. The soul or the sense, i.e. the meaning of the word "Leben" cannot be recovered and also cannot be reproduced with such a method because only the sound or the body of language is detected.

1.2. The Theory of Centers and Plasticity for the Explanation of Language

As already mentioned, scientific analysis of the *activity of the central nervous system*, involved in the physical processes of speech, meets with a great difficulty. This is that the sound of language always has a polar connection with the meaning of language. In order to solve this problem it was thought necessary to assume that entirely definite, closely delineated parts of the cerebral cortex are responsible for the storage of linguistic images, for

developing understanding of language, for actual speech
and the will to speak. The theory based on such localization of individual parts of the language process was known
as the "center theory" (cf. *Clara, Rauber-Kopsch, Rein-
Schneider*) (*Fig. 24*) and generalized for all processes in
the central nervous system (cf. *Rein-Schneider*) (*Fig. 25*).

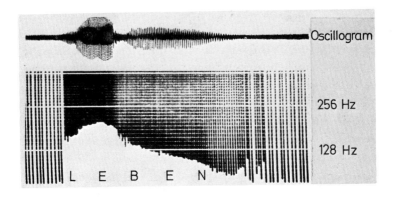

Fig. 23. Course of note amplitude in speaking the word "Leben".
Lower curve: registration with a melody writer; upper curve: oscillogram (after *Grützmacher*, 1938)

An acoustic language center was distinguished from an
optical language center, a reading center from a writing
center etc.; these were all involved in the production
of language and were connected with one another. However,
it was very difficult to achieve a precise localization.
Only the physical processes of "auditory sensation" and
"production of speech" could be localized with relative
precision. Hearing a tone can be demonstrated as an action
potential in the sensory cells of the ear, in the sensory
nerve pathways and finally in the cerebral cortex of the
temporal lobes. This is a purely physical process of excitation, demonstrable at the cortex. The hypothesis was

developed that the primary cortical projection could not be the last stage of the auditory process; on the contrary further areas must be present in the cerebral cortex and involved for example in remembering what had been physically heard. This would be stored like a name written on a piece of paper in a box. The differentiated structure of the cerebral cortex appears particularly suitable for this task. It was believed necessary to assume that memory or an excitation scheme adequate for memory, which could be used at will in roughly the same way as a note taken out from a box, is located in this center. This appeared to be necessary if the word on the notepaper is to be spoken in a sentence. It was thought possible to assume that information containing the necessary word will pass from the "memory center" to an "action center" which would now give an order to the "executive or effector center", e.g. muscles in the larynx are set in motion by the latter center. This assumption was made in order to have a strictly causal chain for the process of hearing and speech.

Because the scheme for "explanation" of language given here was not satisfactory, it was thought necessary to assume that further centers must control the "memory and action centers". Pure "cognition", i.e. finding new concepts, and also processes involving "higher intentions" and "acts of will" would be prepared in these controlling centers. It was believed necessary to assume that these centers were also localized in the cerebral cortex.

These hypotheses were stimulated and to a certain extent supported by observations on patients with speech disorders. In these patients injuries or growths had been found at particular points in the cerebral cortex; these are taken to be responsible for the various disturbances of speech.

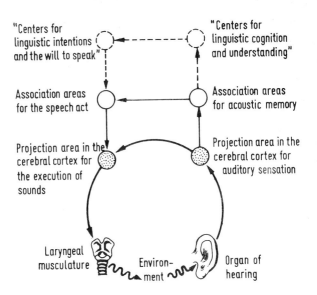

Fig. 24. Organization of audition and speech in the central nervous system according to the Center Theory (*Blasius*, 1967)

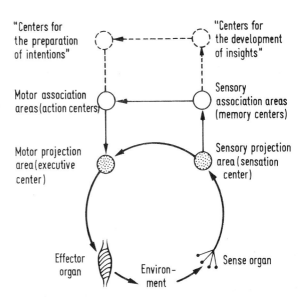

Fig. 25. General organization of the central nervous system according to the Center Theory (*Blasius*, 1967)

These ideas are illustrated by a map of the surface of
the cerebral cortex (*Fig. 26*) in which the position of
the various "sensory association areas" (optical language
center = reading center; acoustic language center = memory
center for auditory images) and the "motor association
areas (motor speech center according to *Broca*; writing
center) are shown.

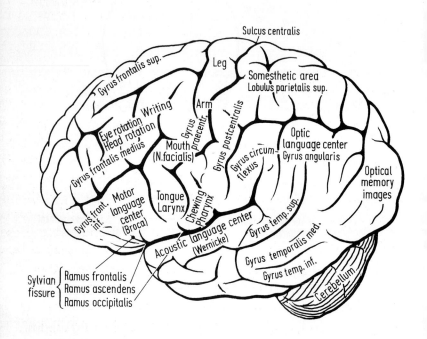

Fig. 26. Motor and sensory cortical fields ("Association Fields",
Ziehen) of the left hemisphere (according to *Rauber-Kopsch*, 1955)

While the sensory and motor centers of the lower level
were designated "projection areas" (*Clara, Glees, Rein-
Schneider*), the expression "association areas" has been
introduced for the centers at the "higher" level.

The principal difficulty, that of localizing "mental activity at a particular point in the cerebrum, i.e. at a "higher" level, has recently led to the opposite theory. This is that such mental activities, e.g. formation of concepts, cannot be localized at all; the brain as a whole must participate. It was supposed that the changeability of the activities requires plasticity in the brain and consequently every part of the brain has the same importance for the whole activity. This line of thinking was summarized as "plasticity theory". Various data found during recent decades are thought to support this theory.

Even 30 years ago resections of the brain had shown that for example the frontal lobes can be almost entirely removed without producing disturbances of speech (*Dandy, Clara*). On the other hand, removal of the occipital lobes is associated with apraxia, i.e. with lasting incapacity to act (*Fig. 27*). If one entire half of the brain is removed, sensory and motor disturbances appear, although these are not so extensive as would be expected from the "center theory". If such an intervention, for example for removal of a tumor, has been undertaken in childhood, then there is no loss of speech (or only a temporary loss). Therefore the "language centers" had not yet been fixed. If the dominant half of the brain (usually the left) or only the *Broca* motor speech center (association area) is removed at a later age, then the capacity of speech could be recovered by systematic training (*Glees*). If one reviews the evidence supporting the "localization theory" on the one hand, and the "plasticity theory" on the other, then it is clear that the sensory and motor projection areas for certain bodily processes can be localized with considerable precision. However, the "association areas" or the "centers for memory images and for activity patterns" can probably never be strictly localized. This result is also supported by EEG data which we shall now discuss further.

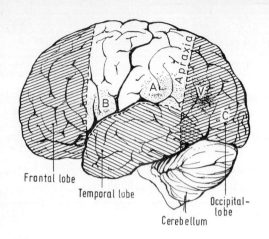

Fig. 27. Schematic representation of those areas of the left cortical hemisphere which can be resectioned in man without severe functional loss (hatched). Resection of the frontal lobe up to the line indicated does not result in any speech disturbance. A = sensory, B = motor, C = visual language center (reading center) (after *Dandy* see *Clara*, 1959)

1.3. Electroencephalographic Results of Excitation of the Cerebral Cortex through Optical and Acoustic Stimuli

As has been known since the investigations of the Jena psychiatrist Hans *Berger* in 1921, action potentials can be recorded from the brain surface and also from the skull surface, which potentials take on a different appearance with changing activity of the cortex, especially with respect to frequency and amplitude. Usually several recordings from the brain or skull surface are made at various points in order to test the activity of the individual areas of the cortex. *Figure 28* shows the effect of an acoustic stimulus of 16 Hz on 6 recordings from the back of the head of a healthy test subject. The initial normal α-rhythm consists of high deflections of average frequency (8-14 Hz) mixed with lower β-waves of higher frequency

(14-20 Hz). It is changed when the tone begins; above all, the α-waves are suppressed. It is noteworthy that this suppression of α-activity varies in the different recordings. If the 16-Hz-tone is replaced by a 1600-Hz-tone i.e. by a tone with a frequency 100 times greater, then almost exclusively β-waves are left. When the tone stops, the normal action potentials of the cortex quickly return.

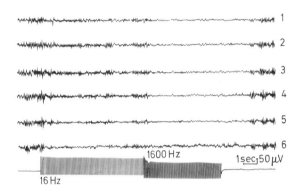

Fig. 28. EEG with an acoustic stimulus of a 16 Hz note and change to a note of 1600 Hz (6 recordings from the skull surface). Notice the different blockade of the α-activity (after *Kugler*, 1966)

It is clear that hearing notes is expressed in a change in the activity of the cortex. However, the kind of change in cortical excitation does not provide direct information on the kind of acoustic stimulus. By the way, this peculiarity is generally valid for the organism in contrast to extraorganismic systems, for which the "law of strict proportionality between action and reaction" is the rule.

The EEG shows entirely similar results when the eye is stimulated with light-dark stimuli. The α-waves disappear initially when the light stimulus meets the eye with this arrangement, too. However, with further stimulation, adaptation takes place, i.e. the α-waves gradually reappear. By the way, the "sensation" of light also gradually falls off after an initial intense reaction. In this case, too, the quality of the light stimulus, e.g. the color or the shape of the stimulus figures cannot be deduced from the EEG curve. Rather more complicated data were obtained in the following procedure: a film was shown to a healthy test person and the EEG recorded simultaneously (*Kugler*). The most striking observation was that the α-activity was most markedly blocked during the scene which was found to be most nauseating. This experimental result illustrates the observation that the nature of the experience including its affective quality cannot be read off from changes in the EEG.

1.4. Electroencephalographic Observations during Mental Activity

Berger found that purely mental activity (e.g. mental arithmetic) is manifested in the EEG. With this experimental arrangement he observed a suppression of the α-waves and a predominance of the β-activity, i.e. an effect entirely similar to that noted with optical or acoustic stimulation of the test subjects. Of course it cannot be determined from the EEG curve what calculation task has been solved nor whether it was solved correctly. A more recent study of the same activity by *Rohracher* confirmed this.

Incidentally it is remarkable that in such and similar studies the α-waves are most regular if the mental activity is at a very low level (*W.G. Walter*). *Rohracher* concluded from this that the basal activity of the cortical structure is accompanied by the α-waves in the EEG and that this is only the precondition for more complicated

excitation processes. These processes lead to suppression
of the α-waves and allow the β-waves to become more prominent. Normally the β-waves are masked by the α-waves.

It can be clearly concluded from all these EEG data that
both optical and acoustic stimuli and spontaneous mental
activity give rise to quite similar patterns of action
potentials in the cerebral cortex. However, qualitative
differences in the stimulus patterns are not correlated
with the excitation patterns. This is generally to be
expected in biology.

The α-activity is clearly related (in part) to the individual central nervous excitability (*Panse*). The frequency
of the α-waves increases with the amount of movement of
the individual person, as could be demonstrated in studies
on a large number of disturbed and normal children (*Panse*)
W.G. *Walter* further observed that the α-waves have a different form in every human being; their individuality
is comparable to that of the finger print.

2. A Phenomenological View of Language

2.1. Soul and Spirit of Language

The foregoing discussions have provided a brief review
of the physiology of language. It was attempted to represent the results which can be obtained in an analysis of
the processes of hearing and speech. We saw that it is
possible to demonstrate with physiological methods physical processes in hearing and speech and in purely mental activity. However, we noticed at the same time that
o n l y the physical side of the auditory sensation and
speech can be subjected to physiological analysis.

Now we wish to turn to the soul and spirit of language, i.e. attempt to give a phenomenological view of human language. We remind ourselves that according to an interpretation of C.G. *Carus*, the soul is the meaning of the body phenomenon and that the body is the phenomenon of the soul. In analogy to this we can now say: the soul of language is the meaning of speech-sounds and that speech-sound is the bodily phenomenon of the soul of language. According to *Klages*, body and soul form the poles of a continuum. This continuum is so intimate that it is difficult for us to separate it in the mind. The difficulty of separation is the reason for the repeated (though, of course, impossible) attempt to attribute specific physical structures in the brain with definite spiritual properties and mental capacities. Such an attempt must always be condemned to failure. "Psycho-physical parallel", the theory of the connection between body and soul, must also be considered a failure; the soul cannot be "constructed" from the body like a line running parallel to another line.

There are good reasons why we rely on the *Klages* theory of the polarity of the body and the soul, the primal continuum which consciousness and the will are forever trying to separate.
We should now give the fundamental characterizations which *Klages* gave for the body, the soul and the spirit. These also make an essential contribution to the understanding of language as body of language, soul of language and spirit of language.

In the chapter "Soul and Body" of his main philosophical work "Der Geist as Widersacher der Seele" (The Spirit as Antagonist of the Soul, 1954, p. 1014 ff.) *Klages* further develops his thoughts. To facilitate understanding of what he meant by the three realms of the human essence, he later gives analogies valid for the world and supports the analogies with language by a striking example: "If we consider that animal vitality has two poles (a relatively receptive and a relatively executive pole) and if we ask from a psychological viewpoint what corresponds to it in the realm of life, taking first of all that of man, what the body is, what the soul and what the spirit is, then the answer is as follows: the body is the bearer of sensation and spontaneous movement, the soul is the bearer of contemplation and formation (= "weaving"), the spirit is the bearer of acts of intellect and will. The body accordingly corresponds to the physicality of the world or, expressed more abstractly, to its materiality; the embodied soul corresponds to the phenomenon of the characters of the world, growing in content of soul with growing predominance of soul; the localized spirit corresponds to a world with selfidentical objects (cf. *Fig. 29* and *Table 5*). These determinations put us in the position to name with certainty the proportion of the body, the soul, and the spirit involved in any our life processes whether it is capable of being conscious or not capable of being conscious (such as nutrition, growth, mating, respiration, excretion, sleeping, dreaming, waking, sensing, looking, perceiving, remembering, viewing, judging, thinking, wishing, planning, willing, doubting, hoping, fearing, reflex-movement, expression-movement, instinctive action, discretion-movement, etc.)".

We should stop at this essential passage and notice that *Klages* gives with these arguments the concept of an entirely new physiology based on the polarity principle and takes into account in particular the role of the spirit in human vital processes. *Klages* has left us a great task which we should at least try to solve step by step.

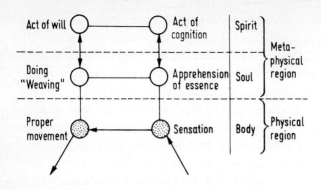

Fig. 29. The human being as triad of body, soul and spirit and their receptive and executive poles (after *Klages*, see *Blasius*, 1967)

Table 5. Body, soul and spirit of man and their equivalences in the world (after *Klages*)

Essence of man	Equivalent in the world	Example
Spirit	World of identical objects	Score
Soul	Appearance of the world characters, growing in psychic content with growing predominance of the soul	Composer who expresses the essence of the world in his music
Body	Physicality or materiality of the world	Music which is reproduced by the human voice or by mean of a musical instrument

Klages further directs his thoughts to human language. I do not believe I can do better than again present his own words:

"Since the capacity to conceptualize is also involved in the most important of all vital processes, namely that of language, the question arises as to what respect it bears witness to the body, the soul, and the spirit.

There is hardly another question which permits us to recognize so directly what research can accomplish with correct metaphysics, and what failures it must suffer if its metaphysics is false. The soul has been losing ground since the Renaissance, in that only the physical was still differentiated from the mental (apart from the exceptions which must be taken into account again and again). Hence linguistics recognized (and even today to a large extent recognizes) only the body of language besides the spirit of language and is involved in many difficulties which basically cannot be resolved and which make all retrospective adaptations illusory. The body of a name is naturally the sound, the spirit is the concept; but to want to understand the genesis, evolution, and essence of language from these alone would resemble attempting to understand a piece of music from the instruments and the score and forgetting the composer! What is the soul of the name? The meaning of the name. We shall mention a few examples of the variety of name meaning and concept.

Everyone knows the difference in meaning of bowl, cup, basin, turreen, sauce bowl, saucepan, pot, tankard, urn mug, amphora, vase, stewpan, bucket, tub and would be able to identify examples of these categories with the help of such knowledge; however, who would be prepared to volunteer to define them abstractly, even with relation to use, in such a way that the definition would exclude confusion!

In several cases, despite their referring to the same concept words diverge to such an extent that it would be impossible to use one word in place of the other. If someone wrote in a death notice instead of "Yesterday evening our dear aunt passed away peacefully" for example "Yesterday evening our aunt croaked peacefully", then justifiable doubt would arise about his mental state. If

one considers the following list of words for dying with
which one conveys respect or disrespect for the dead person and partly a reference to the cause of death: to pass
away, to depart this life, to decease, to expire, to fall
(in war), to perish, to be killed, to pass on, to turn
up one's toes, to kick the bucket, to peg out, to hit the
dust, one will no longer be in danger of confusing concept and meaning! An entertaining joke, which is by the
way very instructive, is the following description: "He
killed himself, sank soulless to the ground, and gave up
his spirit". Linguistics has not missed this. There has
been a great deal of brilliant writing around this, but
it is entangled in the misleading opposition of body and
spirit. The difference in meaning of terminologically
identical names has been subjectively interpreted by hoping to find the reasons for them in feelings which are
supposed to accompany the understanding as well as the
speaking of the word. On the contrary, it is really the
meaning to which feelings are attached; these feelings
are, by the way, very different in intensity and kind
from person to person! It is not our task here to show
how many quite erroneous views necessarily arose from such
thinking".

2.2. The Contrast between Linguistic Meanings and Linguistic Concepts

In order to clarify the part played by soul and spirit
in language, *Klages* gives in a few brief points a quite
excellent review of the contrast between the meaning and
concept of language. I shall now follow this review with
short commentary because the essentials of *Klages*' ideas
are expressed in this summary and at the same time can
be exemplified (see *Table 6*).

Table 6. Distinction between linguistic meaning and linguistic concept (after *Klages*)

		Linguistic Meaning (= soul of language)	Linguistic Concept (= spirit of language)
1	Description	The meaning has the same relationship to the name as the soul to the body	The concept is a meaning which is exhausted through limitation
2		The meaning is the content of language (which is never lacking)	The concept is a dispensable artificial product without which language remains understandable
3	Age	Meanings are as old as language	Concepts have existed only since the beginning of the historical period of mankind
4	Origin and Development	Meanings change like names, but faster as those	Concepts are found at some time and then replaced by other concepts; however, they do not change
5	Definition	Meanings are experienced and hence not definable	Concepts are thought and hence always definable
6	Content	Meaning of names are inexhaustible like the soul	Concepts as such are void
7	Mutability	Meaning of names change with their appearance, even if imperceptibly	Concepts always remain the same
8	Translatability into other languages	Meanings of words are difficult to translate into other languages	Concepts can be precisely translated into any foreign languages of the same historical age
9	Sense and purpose	Meanings serve indicative thinking	Concepts serve comprehensive, discriminatory thinking
10	Character	Meanings have the character of distance	Concepts have the character of nearness

The meaning belongs to the word as the soul to the body; the concept is limited by the delimitation of the meaning.

The first sentence utilizes the polarity of soul and body in relationship to meaning and name. The inherence of soul in a body is taken as an analogy to that of meaning in a name.

An appropriate definition of a concept is moreover "Entschöpfung" (by the way a new word from *Klages*) by delimitation of meaning; the meaning is restricted by conceptual fixation and "entschöpft", i.e. the soul is depleted in this way.

The meaning can never be lacking; the concept is an artificial product, absence of which does not change in the least the comprehensibility of language.

This second statement clarifies the first and says additionally that the comprehensibility of language is linked only to the meaning of what is spoken. At another passage *Klages* illustrates this by the following statement: "Every linguistic statement originally discloses the character of the phenomena and does not serve thinking".

Meanings are as old as language itself, but concepts are not older than historical humanity (and even younger).

Klages says here very clearly that human language has had a soul and a being since its genesis, but that concepts (the *spirit* of language) only arose with the history of man.

Meanings and names change, but meanings more quickly, as the soul changes more quickly than the animate body; concepts are at some time found at first and must sometimes make place for new ones, but they do not change.

Klages' statement that the soul changes more quickly than animate body and therefore the meaning of language more rapidly than names appears remarkable. The modification of the soul and the meaning of language is contrasted by *Klages* with the rigidity of concepts and also with the spirit of language. Meanings are experienced and are consequently not definable; concepts, which are always definable, are thought.

Experience is postulated as the basis of the meaning of language and the defining of thought as precondition for concept.

The meaning of names is inexhaustible, as is the soul; however, the concept as such is without content. Consistent, but surprising is the formulation of the lack of contents of concepts as such. I think this realization is highly illuminating.

The meaning of a name changes, even if sometimes imperceptibly, with its use and is hence never the same in any sentence. It is not even exactly the same in a sentence which is heard and a sentence which is read; the concept is always the same wherever it occurs.

This point underlines the mutability of the meaning of language yet again. It is still traceable in the finest nuance of language. On the other hand, the concept is always the same, provided that is still current and not obsolete.

Meanings of words can never be precisely translated into foreign languages; concepts are absolutely the same in every foreign language in the same or approximately the same historical epoch.

Klages attributed to this the difficulty of translating the meanings of words and the ease of translating verbal concepts, if in the latter case the languages are of the

same historical epoch. This also explains why in general
the translator of a poetic work has greater difficulties
than the translator of a scientific work, for example,
a textbook in physics or physiology.

Meanings, and only meanings, serve indicative thinking;
concepts serve conceptual thinking i.e. the capacity to
form judgements.

Klages distinguishes indicative thinking which is concerned with the meaning of speech content, from discriminatory thinking, which uses definite concepts. We remember
from *Klages*' theory that essence and soul are incomprehensible and that they can only be indicated.

2.3. The Difference between Meaning and Conception Words

Klages' consideration of the meaning of words and the
difference between meaning and conception words is highly
illuminating, and will be quoted without abridgement:
"Only the sentence structure decides the respective intensity, depth, scope, color and power of the word meaning,
there are certainly "conceptually weak meaning words"
and "conception words with a low meaning content" in every
language which has had a long philosophical tradition.
We have used these phrases earlier but we are only now
in a position to understand and explain their full meaning. Since "the object of thought pure and simple" is
what we understand of a matter and mean by it, objects
of thought are concepts and as such not definable in terms
of space and time. If there were names of which the function would be exhausted in defining conceptions, they are
in no way different from the essential conventional symbols of mathematics. Even more so the favorite wish of
the Baroque period, which was to discover a logical al-

gorithm (number language) would no longer be utopian: the names in question would precisely designate objects of thought outside the limits of space and time without the restriction from the spatial and temporal movement rhythm of language. Such names do not exist and will not exist as long as people communicate with language; the meaning content of every name fluctuates to an extraordinary extent, depending on how the mode of description endeavors to express it (whether successfully or unsuccessfully) or whether on the contrary only the concept to which the name is attached is communicated. Such an effort, which gives the scientific treatment of language its sober character, is soon more or less supported by the name itself. If the reader reviews in his mind the most important terms in the last chapter for main or key concepts, he will also think only of the theory of concepts which have dealt with such as: the material, the thing, the resistance, the point of disturbance, the unity, the one, the being. The reader will not hesitate to admit that they themselves are already as plain as science requires when compared with words such as: the dawn, the winter, prehistory, the sea, the climate, the flame, the firmament, the storm, the malediction, sin, misery, ruin. These first are very convincing examples of "conception words with a low meaning content".

In conclusion, *Klages* mentions the nearness character of conception words and the distance character of meaning words. This section constitutes a wonderful sample of *Klages'* deep reflection on the essentials of language and his striking psychological interpretation:

"Whosoever has only a little practice in catching a glimpse of meanings and paying attention to attitudes with which the soul becomes certain of these meanings, that person cannot entirely fail to notice that the conception words

which have a low meaning content are more or less lacking
in distance (while, in contrast, the others never). The
meaning content which still remains now becomes palpably
nearer! However, since the object of thought has relation
neither to distance or to nearness, since it is outside
space and time and can be "touched" hardly at all, the
cause of this must lie in the vital events which have made
it possible to "entschöpfen" the concept of the meaning
of the name. And thus the almost embarrassing nearness
character which is part of the meaning content of every
name without exception is given to us.

The name was either in long use in the service of conceptualizing thought or served to clarify communication
techniques of verbally expressing practical reason clearly for the mere labeling of concepts. This confirms that
we must seek in the here and now the resistance which
we find so disturbing as the source of all concepts and
of conceptualizing itself. Whether one calculates a light
refraction angle or ascends Platonic dialectic obelisks,
whether one pierces infusoria or traps nebulous spots of
the universe on light-sensitive plates, whether one ana-
lyzes the concept of a triangle or the concept of God:
his thinking is blind to distance and resembles a well-
honed knife which is useful for precise sectioning into
smaller and smaller slices. The more successful such
thinking, the greater its flatness. The materialist seeks
his pride in this, the ideologist hides it from himself;
but his deluding himself is in vain. One studies which-
ever idealistic system one wishes. Even if it is the most
important that we know, the philosophy of *Plato*, the meaning content is tested against the psychical aftertaste.
One knows that it is just as lacking in atmosphere and
devoid of a depth as any atomistic theory or a Newtonian
scientific calculation. However, to the extent that such
an aftertaste judges differently, one will inevitably

find that the object of thought had transiently slipped away for the materialist as for the ideologist because the never apprehensible soul of the language asserted her right to flood the reason of the thinker with the mystical abundance of meaning.

But yet another piece of evidence comes to us unsummoned. Just as finding the *existent thing* constitutes the fundamental first step of mental experience, finding the will of the ego constitutes the fundamental second step. It was one and the same primal point of disturbance, namely the Here which is doubled in meeting resistance, from which we would have to start in order to determine the vital grounds of enabling both the existent thing and the *will of the ego* to be experienced is accordingly shown that something without essence for which we thank the Danaen gift of consciousness of existence that our spiritual ego obtrusively approaches, so long as our spirit is awake but not occupied in another way. The nearness of the object of thought corresponds to the lack of distance of the personal ego in us, the victory of the distancing images over concepts and things in us corresponds to an excess of physical enthusiasm or passion, but always to loss of self-*awareness*. Anyone who immerses himself in *Eichendorff*'s stanza which begins with the verse "The distance discourses in ecstasy" does not "*think*" the distance, but has experienced it; and anyone who now reflects on what he has experienced and tried to "think" the distance would have substituted in no time at all something which has a given distance. However, it shares with the small and smallest distance the nearness of all objects of thought and conceals the character of distance to no greater extent than for example the face of a pocket watch.

Not everyone can exchange the world of conceptual content, which is so customary for us, with the world of meaning content, or exchange one-sided rational thinking with philosophical immersion. If the poet - we mean the true poet - temporarily deludes the spirit reflecting in two dimensional terms, the magic of language helps him in consoling the listener and reader with the entirely false, but with the superficially natural conviction, that the content of a poem is not reality. The advocate of deep *meditation* has a far more difficult task because in meditating and speaking he can only *indicate* an incomprehensible and at the same time demands recognition that just this is reality! He may consequently only invite those persons to accompany him on such paths to whom truth does not appear to be paid for too dearly with the destruction of all the opinions with which the strongest driving force of historical man, self assertion, has shut itself off and fortified itself against the power of the will-killing view of the distance".

The phenomenological side of comparison is closed with this pregnant and profound characterization of human language by *Klages*. If I have quoted from *Klages*' works in great detail and have given little commentary, then this is an indication of my conviction that the psychological discoveries of *Klages* and the profundity of his thinking cannot easily be enhanced.

3. Summary

In conclusion we summarize the knowledge and insight gaine from the comparison of physiological and phenomenological data on the living phenomenon of human language.

1. A purely physiological analysis of language can only yield reductionist statements on the genesis and course of bodily processes, such as audition or speech, because of its natural scientific premises and methods.

2. It is entirely possible to describe a causal chain of processes taking place in the temporally proceeding excitations in the sense organs of hearing and speaking, in the related nerves and in the central nervous system, particularly in the cerebral cortex.

3. While certain cortical areas which convey auditory sensation and vocalization can be spatially localized, it is not possible to determine precisely the spatial localization of "association areas" for "memory of auditory images" and "formation of linguistic concepts".

4. The "center theory" is restricted by the impossibility of demonstrating psychic and spiritual capacities at a specific point in the body with the methods of natural science.

5. In place of the inadequate natural scientific interpretation of human language which results from the false assumption of a antithesis of body and spirit stands *Klages*' theory of the trinity of the human essence according to body, soul and spirit. This theory satisfactorily specifies the living phenomenon of language according to its essence, its meaning and its conceptual possibilities.

6. On the basis of the polarity which is observed in the whole cosmos, according to *Klages*' teaching body and soul stand in a polar connection. However, the spirit, which with respect to the "receiving" side of the human essence shows itself as a cognitive act, and in respect to its "acting" side as an act of will is opposed to the polarity of body and soul. *Klages* designated the spirit as the cause of the disturbance of the "contemplating" soul and the "acting" body.

7. According to *Klages* the spirit is outside of space and time and cannot be spatially localized. On the other hand, the soul is involved as contemplating soul in every bodily sensation and as an active or moving soul in every movement of the body.

8. In analogy to these determinations of body, soul and spirit, the meaning of linguistic sound can be referred to as the soul of language and the linguistic sound as the body of language.

9. *Klages*' distinction between the soul of language and spirit of language also allows the meaning of language and the concept of language to be clearly separated from one another and to be described.

10. In his main philosophical work *Klages* shows that a theory of human life should only be constructed on the trinity of its essence of body, soul and spirit. In this sense, the contrasting of physiological and phenomenological results in the explanation of human language can be considered a beginning.

4. Postscript

Many critical voices which are raised today on the situation of man mention the human language which is threatened by decay and destruction and are consequently concerned to revive and renew it. Those who have read the discussions by *Klages* on the essence of language and inwardly digested these can be in no doubt that a renewal of our world which appears more and more to have come under the sole control of that part of mankind chiefly steered by will can only succeed by consideration on the living and figurative content of language. Such a reflection is a renewal of the soul as *Klages* wished from the depth of his heart and love of life.

V. Stereoscopic Vision and Color Discrimination: Their Typological Polarity and Relations to Pictorial Creativeness

> For the eye, pictorial art is more
> real than reality itself. It esta-
> blishes what man would like and ought
> to see, not what he usually sees.
> *Goethe*, The Eye
>
> If the eye were not like a sun, it
> could never glimpse the sun. If God's
> own power did not reside in us, how
> could we be enraptured by the divine
> essence?
> *Goethe*, Zahme Xenien

Form and color are the two essential, clearly distinct qualities of visual perception. We can further distinguish two different aspects of visual perception of both form and color, a somatic or physical aspect, which may be termed "sensation" (in German: Empfindung), and a psychic aspect, which we may call "real perception" or better: "true vision" or contemplation (Schauung). Contemplation, which according to *Klages* stands in polarity to sensation in the same way as the soul to the body is not to be confused with the conscious act of apprehension, which relates to the conceptual isolation of certain forms or colors in visual experience. In contemplation, the totality of visual experience is connected to former experiences of a similar kind and evaluated comparatively. Sensation is dependent on sensation; for there is a form of contemplation, which does not involve sensation, such as that which is possible in dreams, or in day dreams, when the eyes are closed. Contemplation is then the essential metaphysical event within visual experience. In a conscious act of apprehension only those sense data are received which can be ordered to the contemplation, that is to say, to the image in the soul. In the act of apprehension, by contrast, one becomes conscious of those data which are incompatible with contemplation or which draw the attention to particular details.

Physiological methods permit the sensory data perceived and consciously apprehended to analyze in view of the type, dimensions and accuracy of sensation and also to determine the amount of time required for the sensory process.

On the other hand, it is possible to gain insights into the psychic aspects of the processes of perception, by means of psychological tests, in other words, into contemplation, and the formative processes related to it.

Extensive experiments carried out with my colleagues R. *Repges* and L. *Ziegenhain* were designed to throw light on the connection between the perception of form and that of color in visual experience. Because of the novel methods and results, a more detailed description is necessary here. At the same time, this introduces important problems of research into life.

1. Physiological Methods

1.1. Testing Stereoscopic Vision

Conscious apprehension of spatial form is termed "capacity for spatial vision" or stereoscopic vision. For the examination and evaluation of this capacity we chose the method of Eb. *Koch* with which I had already previously obtained excellent results. This constitutes a refined quantitatively evaluable procedure for testing the capacity to perceive depth (*Koch*, 1941). Koch's apparatus for testing spatial vision is based on the possibility (first indicated by *Rollmann*, 1853) of viewing in depth dichromatic stereoscopic pictures through dichromatic glass discs. In front of a ground glass disc evenly illuminated from behind, half images of a three dimensional object, for example a pyramid with varying lateral distances between the apices, can be placed in this apparatus, where they may be exchanged at will. These stereoscopic pictures are so drawn that the picture on the left represents a pyramid outline with red contours seen from above, and that on the right one with green contours. Two windows are built into the front of the apparatus, with green glass on the right and red glass on the left. On looking at the stereo images through these colored windows, the observer gains a stereoscopic impression because only the left red contours of the stereo picture pass through the green glass on the right and only the right green contours are visible through the red glass on the left. The test subject adjusts a moveable pointer to the point of the physically seen pyramid, because the two half images of the pyramid are fused centrally (i.e. in the visual cortex) into a stereoscopic image. The distance of the pointer from the level of the picture can then be read off in mm on a scale (*Fig. 30*).

Simple calculations show that the height of the pyramid h which is to be adjusted is determined by the distance from the eye A, the distance of the picture E and the lateral distances of the half images a. This is given by the following equation:

$$h = \frac{a}{a + A} \cdot E \qquad (1)$$

With constant E and a given eye distance A, h is a hyperbolic function of the lateral distances of the half images a.

The degree of deviation in the adjusted value from the calculated value for the height of the pyramid is a measure of the accuracy of stereoscopic vision. The extent of the depth perception capacity a_{max} can be determined by stepwise increases in the lateral distances of the half images a - and thus the lateral disparity - up to the limit of the possibility of fusing the images. Moreover, with a specific value of lateral distance one can measure the time required for three adjustments to the top of the pyramid.

In order to be able to compare the performances of different subjects with one another, the following statistical calculations for the accuracy, the extent of stereoscopic vision and the time required were carried out.

A certain deviation (in mm) with a small lateral distance of the half images has a larger weighting than such deviation with a large lateral distance because of the hyperbolic course of the expected values (cf. *Fig. 30*). The deviations of the distance in mm between the expected values for the eye distances from 50 to 75 mm were therefore chosen as a reference value. This distance h'_A becomes greater with increasing lateral distance in accordance with the course of the curve. The value of the height which is adjusted is divided by the respective h'_A for the adjusted lateral distance:

$$h_{corr} = \frac{h'}{h'_A} \qquad (2)$$

The sum of the deviation from expected values of height h_{corr} collected in this way divided by the number of measurements N gave the total variance. The reciprocal value of the total variance was named the accuracy G:

$$G = \frac{N}{10 \cdot \Sigma (h_{corr})} \qquad (3)$$

From the value for the accuracy G, the values measured for the extent a_{max} and the time required for the adjustment t, the index R was calculated:

$$R = \frac{G \cdot a_{max}}{t} \qquad (4)$$

The index R for each subject was divided by the average value $\bar{R}$ of all the subjects to obtain a value X_R, the relative performance index for stereoscopic vision:

$$X_R = \frac{R}{\bar{R}} \qquad (5)$$

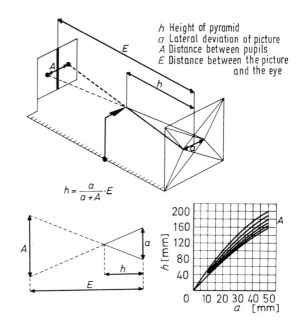

Fig. 30. Principle and evaluation of the investigation of stereoscopic vision after *Eb. Koch* (see *Blasius*, 1970)

1.2. Testing the Capacity to Discriminate Color

In collaboration with L. *Ziegenhain* a new procedure was developed to test the capacity for color discrimination. The subject is required to arrange the 48 colors of the *Ostwald* color disc (*Ostwald*, 1923). Tested were number of errors, the number of deviating color regions and the time required to perform the test. The performance in these three tests was likewise calculated and arithmetically summarized in an index Y_F.

The 48 pure colors of the *Ostwald* disk contain neither white nor black admixtures. The color tests of the 48 color shades were equal in area and each stuck on to a card; they could be assembled into a color disk in which one test color followed another with a similar shade and each color stands opposite a corresponding complementary color which can be mixed with it to form white or neutral grey (cf. *Blasius, W.:* Arch. Psych. 122, 67-91 (1970), Table 1, Fig. 3).

The studies were always carried out under the same conditions of illumination.

The test subject had to solve the following three tasks:
1. The color test cards had to be arranged in a color disk and placed in the appropriate slot, so that each color card followed the one with a neighboring color. The 48 color shades were already divided into 4 regions and the starting color for each region was already marked. The total time required to solve the task was determined.

2. All the 48 color tests were to be arranged as quickly as possible without a previous labelling of color values. The trial was always stopped after two minutes and the results determined.

3. All 48 color specimens were to be arranged, and the sequence could not be changed after placing.

The number of mistakes in all three tests, the extent of confusion, i.e. whether neighboring color shades or those which lay far apart had been confused, the color region in which mistakes were made and the total time required for the arrangement were registered. The value of the index Y_F was then calculated for each test subject.

2. Psychological Methods

To obtain data on the perceptual capacity and creative ability of a test subject the following procedures have proved particularly useful: 1. investigation of color selection (a method developed from the test for ability to distinguish color), 2. the creativity test according to *Lamparter*, 3. the self diagnosis test modified according to the suggestions of *Scholl* and *Vollmer*, and finally 4. the type diagnosis according to *Kretschmer*, which was supported by photographs of the test subjects in two planes.

2.1. Investigation of Color Selection

Subsequent to the color discrimination tests a color selection test was carried out by the test subjects. From the 48 specimens of the color disk the test subjects selected their favorite color and a color matching this as well as three or four other colors harmonizing with each other. It was observed above all whether complementary, i.e. colors from the opposite side of the disk, or neighboring colors were selected.

In the selection of the two colors a nominal distance of 24 colors of the 48 color disk was assumed. The deviation

D_1 from this expected value was counted and divided by the average value $\bar{D}_1$ of all subjects for calculation of the index Z_{2-F}:

$$Z_{2-F} = \frac{D_1}{\bar{D}_1} \tag{6}$$

The higher the value Z_{2-F}, the smaller the extent to which the colors in the color disk lay next to each other. The value of Z_{2-F} was zero when the colors selected were complementary.

2.2. The Creativity Test

In order to gain an insight into the kind of creativity shown by the test subjects, they were asked (according to the suggestion of *Lamparter*) "to paint a carpet just as they wished". Water colors and opaque colors, water-color paper and drawing paper as well as thick and fine brushes were made available. This task is little restricted with respect to technique and content so that important conclusions as to the pictorial creativity and the type category can be made from the type of painting or drawing, the content and form of the picture, the color selection and arrangement.

2.3. The Self-Diagnosis Test

At the end of the study the test subjects were given a questionnaire for self diagnosis. This contained five questions suitable for type diagnosis from the *Scholl* questionnaire as well as five further questions about the kinds of art preferred.

The way in which the ten alternative questions were answered depended on the preponderance of schizothymic or cyclothymic attributes.

There were four possible ways of answering:
a) decision for one of the two properties or tendencies (value 1.00),
b) pre-dominance of one of the two attributes or tendencies (value 0.75),
c) joint presence of the two attributes (value 0.50),
d) not answerable (value 0).

2.4. The Experimental Material

The studies were undertaken in 30 test subjects, 17 males and 13 females, aged between 15 and 40 years old. All the test subjects had volunteered for the study; no special selection procedure was followed. The test subjects were chiefly students of medicine, psychology and educational sciences, physicians, salespeople and technicians, apprentices and schoolchildren. The test subjects were not informed in advance about the purpose of the study.

3. Performance Differences in Stereoscopic Vision

Testing the stereoscopic vision with respect to the extent, precision, the time required and the performance index X_R, calculated from 30 test subjects showed that the height of the index is generally parallel to the extent of depth perception, with its precision and with the time required.

Moreover it is striking that depth perception ability shows a great scatter in performance range (in the present test material this is about 1 to 67). Consequently much greater differentiation is possible than the usual previous procedures.

4. Performance Differences in Color Discrimination

Testing the color discrimination in the 30 test subjects with the new procedure showed clearly that the performance range of color discrimination is also great and that it is about 1 to 29 for the criteria chosen.

5. Typological and Sex Differences in the Kind of Creativity

Assessment of the creativity task according to *Lamparter* (1932) enabled the test subjects to be allocated to the "form-preference" or "color-preference" group. According to the studies of *Lamparter* and the earlier studies of *Blasius*, the form-preferrers are characterized by drawing their "carpet" on hard paper with a fine thin brush. They prefer a dry painting technique and use opaque color and little water. The pattern is often first pre-drawn with a pencil; great value is laid on the exactness of the lines and the form. The design is dominated by geometric figures, circles, squares, rhombi or symmetrical arrangements, also snake or Mäander patterns. These designs are evenly painted in strong color shades standing out clearly from one another in contiguous areas, mostly with complementary colors; every area is often edged with stron black or a dark shade in order to mark it off clearly from the neighboring areas.

Form-preferrers seldom try to attain picturesque effects by mixtures of colors or allowing colors to flow into one another. The color serves the form-preferrer only as a means of emphasizing the form; the elements of form determine the composition.

The color-preferrer solve their creativity task entirely differently. They use chiefly soft absorbent paper, a thicker brush and water colors which are put on the paper as fluids with a great deal of water. They paint with much colors but do not separate these, allowing them to flow into one another, so that many color transitions and color nuances arise. They rarely use the given color shades but produce new shades for themselves by mixing. Form elements are rarely present in their carpet designs or are only hinted at.

There are transitions between the ways of the form-preferrers and the color-preferrers paint, so that not all subjects can be categorized in the one type or the other. Hence two further groups of form-color preferrers and color-form preferrers were formed; into these the other cases were allotted according to the preponderance of their form or color preference.

Of the 30 test subjects 5 were classified as pure form-preferrers (group I), 7 as pure color-preferrers (group IV), 10 as form-color preferrers (group II) and 6 as color-form preferrers (group III); 2 test subjects could not be unequivocally classified.

The breakdown of the female subjects was interesting: 4 color-preferrers, 4 color-form preferrers, 5 form-color preferrers. Remarkably, none of the female test subjects could be classified in the group of pure form-preferrers (cf. *Table 7*).

Table 7. Indices of Stereoscopic vision and Color discrimination arranged in creativity groups

Experimental subject	Sex	Creativity group	Stereoscopic vision		Color discrimination	
			X_R	Average	Y_F	Average
Ku.	m.	I Form preferrers	5,960	2,689	0,089	0,514
Wa.	m.		3,785		0,765	
He.	m.		2,759		1,278	
Be.	m.		0,563		0,413	
Bl.	m.		0,378		0,125	
Ha.	f.	II Form-color preferrers	2,535	1,311	1,250	0,926
Vi.	m.		1,811		0,250	
Wit.	f.		1,538		1,087	
Sw.	f.		1,139		0,500	
Sm.	m.		0,998		1,932	
Win.	f.		0,612		0,994	
Te.	f.		0,547		0,667	
Wie.	m.	III Color-form preferrers	0,876	0,362	0,712	1,006
Wis.	m.		0,623		2,000	
Rü.	m.		0,468		1,540	
Pl.	m.		0,436		0,313	
Me.	f.		0,338		0,111	
Kü.	m.		0,245		0,424	
Ho.	f.		0,231		1,665	
Un.	f.		0,197		1,525	
We.	m.		0,109		1,432	
Os.	f.		0,089		0,334	
Ja.	f.	IV Color preferrers	0,543	0,276	2,540	1,506
Wel.	f.		0,450		1,000	
Gö.	f.		0,285		1,540	
Sa.	m.		0,151		1,430	
Gu.	m.		0,119		1,665	
WiW.	f.		0,109		0,846	

6. Stereoscopic Vision and Color Discrimination in Relation to Kind of Creativity

A very informative result was found by correlating the indices of stereoscopic vision and color discrimination of all the test subjects with the results of their creativity tests. In Table 7 the values of the indices X_R and Y_F are assembled, arranged according to the creativity groups I - IV. It can easily be seen that the indices for stereoscopic vision X_R in the form-preferrers are highest both in absolute terms and on average; on the other hand the indices of color discrimination capacity Y_F of the same subjects are the lowest on average. The color-preferrers show the converse behavior. Their index for X_R is the lowest on average and that for Y_F is the highest. The groups of form-color preferrers and color-form preferrers stand in the middle between the two other groups in this respect (cf. also *Figs. 31a* and *b*).

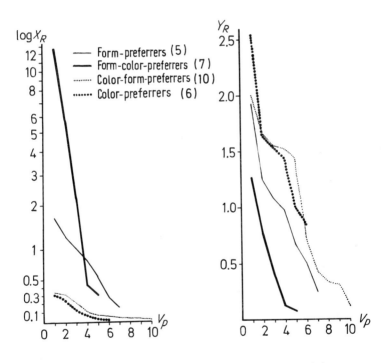

Fig. 31. Performance indices of stereoscopic vision X_R (a) and color discrimination Y_F (b) arranged according to performance in the four groups of creativity types (*Blasius*, 1970)

7. The Complementary Behavior of Stereoscopic Visual Capacity and Color Discrimination Ability

Since pronounced stereoscopic visual capacity is associated with only a low ability for color discrimination and vice versa, there is a complementarity between these two capacities. A graphical representation of the indices for X_R and Y_F of all subjects shows this very clearly (*Fig. 32*); by means of three centers of gravity a hyperbola can be constructed through the values; this expresses the average tendency. There is a large number of subjects in the area of moderate to low performance with respect to both capacities. On the other hand a highly pronounced development of one of the two capacities is accompanied by a loss of the other.

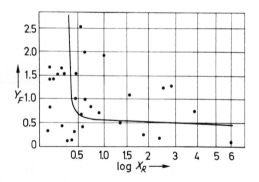

Fig. 32. Correlation of the indices of color discrimination Y_F and stereoscopic vision X_R (*Blasius*, 1970)

Since the two indices, that of stereoscopic vision X_R and that of color discrimination Y_F, are each calculated from three comparable parameters, it appeared worthwile to correlate these parameters separately with each other.

The two capacities are each integrated for accuracy with respect to the number of mistakes, the extent of the performance range, and the time required respectively.

In order to be able to compare the individual components of the indices X_R and Y_F, a correlation matrix was established in which the correlation coefficient r was determined for every combination of factors. *Table 8* shows the values calculated. The closest correlation, which is expressed by the highest correlation coefficient r, is between the extent of stereoscopic vision and its accuracy (r = 0,921).

Table 8. Correlational matrix of data from the experiments on stereoscopic vision and color discrimination

a_{max}	G	t	B	F_1	T_1	
	0,921	0,809	0,561	0,473	0,921	a_{max}
		0,695	0,602	0,529	0,897	G
			0,544	0,498	0,892	t
				0,791	0,618	B
					0,638	F_1
Correlation coefficient r						T_1

a_{max} : extent of stereoscopic vision (maximal distance)
G : accuracy of stereoscopic vision
t : latency for 3 tests on the stereoscopic visual apparatus
B : number of areas in which deviations were noted during test 1 on color discrimination
F_1 : number of errors per area
T_1 : total time for test 1 on color discrimination

Moreover, there is a strict relation between the extent of stereoscopic vision and the time required. These correlations confirm the results of earlier studies.

Furthermore, it is significant that for the color discrimination there is a close relation between the extra time required and the extent and the accuracy of stereoscopic vision and its time requirement. The time required is the most important factor in assessing color discrimination capacity.

8. Color Discrimination, Stereoscopic Vision and the Kind of Creativity in Relation to Color Selection

The index Z_W for selection of color was calculated as average value of the three indices Z_{2-F} for the 2-color selection, Z_{3-F} and Z_{4-F} for the 3- and 4-color selection, respectively. The higher the value for Z_W, the greater the deviation from the complementarity of the colors, i.e. the more generously are neighboring colors in the color disk selected.

It is informative to compare the values Z_W for the color selection test with results of the stereoscopic vision and color discrimination tests (*Fig. 33*).

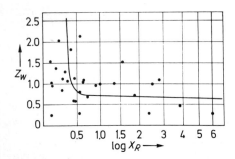

Fig. 33. Correlation of the indices of color selection Z_W and stereoscopic vision X_R (*Blasius*, 1970)

If the index Z_W is represented in relation to the logarithm of the efficiency of stereoscopic vision X_R, then there is a converse proportionality which approximates a hyperbola. This behavior signifies that the preference for neighboring colors is associated with a poorly developed stereoscopic vision, while all those test persons who selected complementary colors showed the best performance in stereoscopic vision.

There is a direct proportionality between the relationship of the color selection index and the color discrimination capacity. The more pronounced the color discrimination capacity, the more frequent the selection of neighboring colors.

If the indices of the individual color selection tests are compared with the index Y_F for color discrimination ability, there is the following result: there is a weak correlation between the 2-color selection (index Z_{2-F}) and the color discrimination capacity (correlation coefficient $r = 0.542$). This is followed by the 3-color selection with the coefficient $r = 0.691$ and the 4-color selection ($r = 0.760$). The strictest correlation ($r = 0.786$) is found for the relationship between the index Z_W for the overall results of the color selection and the index Y_F.

It is furthermore very interesting that the colors preferentially used in the creativity test are the same as those in the color selection test. If one arranges the indices of color selection of each of the four groups of creativity type according to its size, it is shown that color-preferrers select primarily neighboring colors while form- and form-color preferrers mainly select complementary colors.

9. On the Polarity of Form and Color Perception

In order to do justice to the results on differences in performance in the perception of form and color from a physiological and psychological point of view, it is necessary to look once more at the differentiation of sensation on the one hand and perception (or Schauung) on the other.

Visual perception is distinguished on the physical side as "sensation" and on the side of soul as "Schauung"; these constitute a polar relationship of materiality and figurativeness. Thus sensation is experience of the materiality of the real world; it conveys the kind and intensity of the sensory data such as color, degree of saturation, degree of lightness, depth of focus, etc. which are to be perceived. Schauung or the experience of meaning, the character and the figurativeness of reality stands in a polar connection to sensation.

We know from experience that not everything in the continuous flow of sensory experience is registered; only a part of what is seen is consciously perceived. "Conscious perception" is designated the "Auffassungsakt" or "act of cognition".

The continuous flow of sensory experience is divided by incisive cognitive acts which break up this process; these bring the impression of what has been seen into consciousness (*Palágyi*, 1925). What is apprehended by these discontinuous acts of consciousness are fragments torn out of the totality of experience (*Klages*, 1955). In no way it is possible to bring the totality of the impressions entirely into *consciousness*.

By conscious or unconscious direction of attention, parts
of the total impression (which are experienced in temporal succession) can be detached and made to stand out.

The cognitive act can merely bring parts of the stream
of reality to consciousness; the cognitive act distinguishes and separates the experienced images which are
available. The period of time in which what is seen is
apprehended by consciousness is hence of crucial significance for the quality of what is experienced. The longer
and more often the cognitive act makes experience conscious, the more can be apprehended of the totality of
the impression.

Consciousness disrobes the living reality of experienced
images of their essence by stopping and fixating ever
changing phenomena. Since continuity and mutability are
characteristics of real life, sensory experiences also
change with every new moment. The attributes and characteristics of phenomena are fixed and defined in concepts
and unities by conscious activity. Relationships of thing
with one another, forces and causes can only be found and
assembled into systems by fixative characterizations.

Instead of the living reality of events, however, such
systems only represent movements and processes abstracted
from causes and forces. They are therefore laws of natura
processes only insofar as these processes are described
by abstractions (*Klages*, 1955).

The concept is consequently a product of the thinking
spirit; however, the concept refers to a reality which
is not produced by the spirit, since the spirit can only
supply an abstraction of experienced reality. By its content of contemplation, the concept can "remind" one of
something which has been experienced. For example, to

someone who has been born blind the impression of "redness" cannot be conveyed by the term "red", but only to a seeing person who is reminded by the term "redness" of an experience of redness, aided by a living image of the color.

And anyone who has the task of distinguishing colors must judge their equality or inequality by comparing color impressions.

Alternatively, he may have a current sensory experience with which to compare the remembered image of a color he has already seen. Color discrimination is made possible not by stringing together elements of color sensation, but by the connection of image-bearing experience with the reality which gives rise to the images (*Klages*, 1955).

Perception of depth is also built on the experience of space, which actively participates as a memory in the assessment of spatial forms. Only through the connection or polarity of remembrance, retrospective and newly sensed reality it is possible to judge comparison and similarity.

Perception, therefore, derives as it were from proportional encounter of the experiencing person with his whole remembered and experienced environment; it derives from polar tension and not mere opposition. *V. v. Weizsäcker* (1968) expresses this similarity: "Perception is not a replica-like manufactured product, for example, a photograph, but must be viewed as an activity in "becoming". It is not a subjective end product but an accomplished meeting of the poles of living reality".

Perception of color and space is related to sensation of color and sensation of space in a similar way as the soul, character, or the essence of man is related to his bodily manifestation. This soul-body polarity is also the basis of the constitutional types of *Kretschmer* (1967), and the form- and color-preferrers of *Lamparter* (1932) (see page 25). It is the essence of the types that they as *Wellek* says (1966), "live from the polar tension between themselves; they experience their peculiar characterization precisely in this mutual opposedness".

The *schizothymic* type, which according to *Kretschmer* has an excellent faculty for analyzing and abstracting, usually shows above-average performance in depth perception (*Blasius*, 1943). The schizothymic is to be designated as a form-preferrer; his kind of creativity is a formalistic, geometric with "tendency to perseveration", as the studies of *Lamparter* and our own earlier and other studies have shown.

The pyknic *cyclothymic* whose perseveration and abstraction faculty are only developed to a small extent is the polar opposite of the schizothymic (who is physiognomically a leptosomic type); the way of painting of the cyclothymic shows no elements of form or spatial deliniations; his designs are predominated by color nuances and transitions, by diversity and atmosphere. As we have found, the cyclothymic type has an excellent color discrimination capacity.

Finally, the *ixothymic* (who is physiognomically characterized as athletic) has a creativity form between the form-preferrers and the color-preferrers. This type rarely shows pronounced characteristics of one or the other. His viscous, dogged temperament is also intermediate with respect to abstractive and analytic capacity. The athletic ixothymics in our experimental material showed only average performances on the test of stereoscopic vision and on the color discrimination test.

To a certain extent the evaluation of the self-diagnosis questionnaire yielded less satisfactory results. It was shown that the questions on the Scholl questionnaire, which were considered particularly suitable to differentiate schizothymic or cyclothymic properties, were often answered in a cyclothymic direction, while the answers to questions about preferred kinds of art were mostly answered in a schizothymic way.

10. Color Selection as a Polar Phenomenon

Color selection is used in psychology to test mood and temperament. Of course, *Heiss* (1960), *Guilford* (1959) and others emphasize that "a certain color is never to be attributed to a specific experience". The meaning content of a color is consequently not easy to grasp, since a certain color is never experienced alone but always in a particular experiential context. The color impression is almost always combined with other sensory impressions such as that of surface, arrangement, space, and form.

Nevertheless, it cannot be denied that individual colors have a certain psychic effect. *Goethe* emphasized this experiential quality of colors in his color theory (*Goethe* 1827). More recent authors such as *Heiss* (1960), *Hofstätter* and *Lübbert* (1958) also demonstrate these quality impressions linked with certain colors. Contemplation of colors directly conveys their character. *Eysenck* (1941) was thus able to undertake a typing of character attributes from the preference of certain colors.

In color combinations the colors chosen depend on the preference for their component colors. *Allan* and *Guilford* (1936) demonstrated experimentally that either small or great differences in color shade are chosen when several colors are combined.

This observation is consistent with the results of our studies, in which significant differences in color selection could be found between form-preferrers who usually chose complementary colors, and the color-preferrers who chose neighboring colors on the color disk much more frequently. This seems even more remarkable since the color selection could only be made from forty-eight colors in the color disk.

The results also show that a repetition of the color selection test leads to an accentuation of the characteristic differences. It can be further concluded from the study that color selection with combinations of several colors is more suitable for showing typological characteristics than the selection of only two colors, for example. The preference for color combinations of complementary colors or similar color shades is a very persistent characteristic, despite many possibilities that the color selection will be influenced by incidental motives. This is confirmed by comparison of the colors preferred in the color selection test with the colors or color combinations which were most used in the creativity test: the same primary colors and color combinations were often chosen by the test subjects in the two tasks.

It is easier to understand these results if one reflects about the polarity of the perceiving person and the perceived object. The perceiving person does not only sense the shade, saturation and brightness of colors, but at the same time experiences their character and meaning on the basis of the polarity of the perceived object with its image in his soul.

In color selection and formative creativity there grows in the imaginal soul of a person a polar image which only belongs to him. The polarity of experienced image and experiencing imaginal soul is the actual basis enabling perception, as *Klages* recognized. Our own studies confirmed this psychological observation extremely well.

11. Summary

In order to clarify the typological preconditions for form and color perception and color selection, extensive experimental studies with various physiological and psychological methods were carried out.

The investigations yielded the following results: Both stereoscopic vision and color discrimination are developed to differing extents in individual human beings. Typological classification shows clearly that schizothymic persons mostly have higher than average stereoscopic vision and lower than average color discrimination. The converse situation is found in cyclothymics: they are distinguished by an above-average color discrimination and a stereoscopic vision which is little developed. The ixothymics have average performances in both perceptual capacities. There is a complementarity between the two sensory faculties.

Evaluation of the results of the creativity test allows the test subjects to be classified as belonging to color-preferring or form-preferring groups. The form-preferrers, to which the schizothymic type belongs, mostly design in a formalistic-geometric fashion, with a tendency to perseveration; they chose complementary colors to intensify the form. Polar to the form-preferrers are the color-preferrers who belong to the cyclothymic type; their creations are characterized by color nuances and color transitions; elements of form are unobtrusive or entirely lacking in their designs. It was shown that form-preferrers usually had above-average performances in stereoscopic vision and color-preferrers above-average capacities in color discrimination.

VI. Quantitative Thinking in the Life Research

> The earth as one whole should be considered as *one* treasure; it is from it that economics should be learned.
> Novalis

"Thinking in Numbers" is one of the most astonishing intellectual capacities of man. It was the crucial means of founding and building up an exact science and a theory of the natural and vital processes in all spheres.

In making this observation, we should not omit to mention that from this means of knowledge a kind of compulsion has risen in the course of human development; we believe that we can no longer escape from it.

Thinking in numbers, i.e. thinking in units and quantities determined knowledge of nature even during its origin in the antique period, but it has spread out more and more only since the Renaissance and particularly since the 18th century. It has become a fateful "compulsion to count".

That this "compulsion to count" was viewed very early as determining the destiny of man is shown by the famous copper engraving by *Albrecht Dürer* entitled "Melancholie".

This engraving, which was done in the year 1514, gives a fascinating picture of the essence of the Renaissance and the Post-Scholastic period. The awakening of the natural sciences which *Dürer* experienced and in which he participated (one needs only to think of his "proportion theory" and his work on fortifications) is shown in this picture. It shows the technical aids of the new sciences, instruments for measuring space and time: compasses and clock, instruments for weighing and building: scales, ruler and other devices. In the middle of all these tools and instruments sits a brooding woman, the embodiment of melancholy, with key and purse, "the attributes of power and wealth", as *Dürer* himself comments. In the upper right hand corner of the picture one recognizes on the wall a "magic square" which is so constructed that every series of numbers in the horizontal, vertical and diagonal direction gives the same sum.

In his book, *Giorgio Vasari*, one of the first art historians, a contemporary of *Dürer* and admirer of his engravings, enumerates among the engravings which set the whole world in amazement: "The melancholia with all the instruments, which makes the men and whoever needs them into melancholics". This interpretation by *Vasari* is of course too simple; however, it may betray a presentiment of the consequences of the epoch which was then only beginning.

After this digression we will now consider the fateful consequences of quantitative thinking in natural science and life research.

If one assembles the sum of all important numbers which have been obtained in the course of recent centuries on the basis of physical, chemical, astronomical and geological measurements, then one observes a regular increase of these numbers with time, which is characterized in mathematical language as exponential (*Blasius*, 1966).

This exponential growth, which, by the way, is also found in many organic growth processes in certain phases, differ from linear or additive growth in the rapid acceleration of its rate.

The difference between linear and exponential increase in growth is explained in two graphical representations (*Fig. 34* and *Fig. 35*).

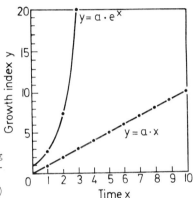

Fig. 34. Additive growth according to the formula $y = a \cdot x$ and exponential growth according to the formula $y = a \cdot e^x$ (*Blasius*, 1970)

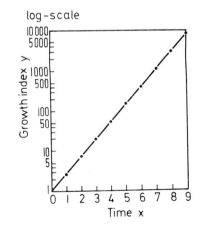

Fig. 35. Representation of the exponential function $y = a \cdot e^x$ in a semilogarithmic coordinate system. (Notice that the steeply climbing curve (see Fig. 34) is transformed into a straight line [*Blasius*, 1970]).

All the numbers resulting from the physical, chemical, astronomical and geological measurements just mentioned are contained in the handbook founded by the chemist H. *Landolt* (1831-1910) and the meteorologist R. L. *Börnstein* (1852-1913). This handbook has been brought up to date in several editions. It would involve an enormous amount of work to assemble the numbers obtained up to a certain point in time. However, for our purposes it is sufficient to determine the number of pages obtained in relation to time (*Fig. 36*). If one selects a logarithmic measure for the page numbers of the handbook and a linear measure for the time (in years), one obtains a straight line representing their reference points. This is evidence that the increase in page numbers of the handbook is strictly exponential. While only a thin volume of about 250 pages could be filled with all the numbers which were important at time of the first edition in the year 1883, the number of pages grows in the following decades and reaches the enormous sum of about 20,000 pages in the year 1964. By extrapolation of the average values for the lower page numbers shown in *Figure 36*, it can be determined that in about the year 1760 all the physical and chemical data which were known could have been assembled on a single page. Research in natural science managed with a very few numbers up to this point. However, it is frightening to extrapolate this function into the future. About the year 2050 a new edition of this work should contain the astronomical number of 1 million pages. If each volume contains 500 pages about 2000 volumes would be expected; these would alone be sufficient to fill a considerable library.

Of course, it is to be assumed that storage and processing of such a huge sum of numbers will be accomplished in the future with the help of computers.

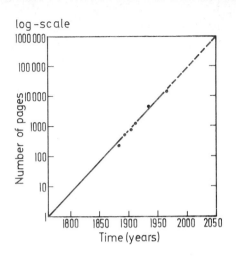

Fig. 36. Development of the total number of pages in the six editions of *Landolt-Börnstein*'s "Numerical Values and Functions from Science and Engineering". Page numbers on a logscale, years on a linear scale. Continuous curve = average values; broken curve = extrapolation (*Blasius*, 1966)

However, the natural scientist does not transform all the phenomena of nature into numbers only in the extraorganismic sphere, but also in the realm of organisms. Measurements which give rise to huge quantities of standard figures are continuously carried out by physiologists and biologists. The National Research Council in the USA has recently published a "Handbook of Biological Data" on which 4041 physiologists and biologists in the western world were collaborators. These together supplied a manuscript of about 20,000 pages which was concentrated into a volume of 584 pages closely printed with numbers in small type. Similar works for individual fields of biology are already complete or are in preparation.

The number of publication organs in which the obtainment or calculation of the data is grounded increases continuously, as can be demonstrated for the special journals in the field of physiology. This increase is also exponential, as proved by corresponding representation of the results from *Rothschuh* and *Schäfer* (*Fig. 37*).

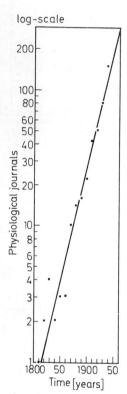

Fig. 37. Increase in physiological journals. Number of specialist journals on logarithmic scale, years on linear scale (modified after *Rothschuh* and *Schäfer*, 1955)

This accumulation of numbers would earlier have been unimaginable. It serves biology, hygiene, medicine, and technology as starting material for their planning and constructions. It had not remained without consequences for mankind. By improving hygiene, technical and medical progress has resulted in an increase in human population which was very slow in earlier millenia but is not precipitous. This was effected by reducing infant mortality and increasing the average life expectancy (*Scheer*, *Blasius*, cf. *Fig. 38*). While only about 10 million people lived on the earth in the year 7000 BC (today so many people live in London, New York or Tokyo alone) and a doubling of this number took place in 2500 years, there were already 100 million people on earth at the time of Christ. However, a further doubling took place in 900 years, and in about 1800 there was a sharp rise in the rate of population increase. In 1850 the total number of human beings had already grown to 1250 millions; according to the most recent censuses there is a total of 3600 million people in the year 1971; the world population now increases by about 186.000 per day and by about 70 million people per year. The increase in the earth population in the last hundred years alone corresponds to the total number of all people who lived in the course of all the millenia of prehistory and all the centuries of history. An extrapolation of the growth curve shows that 5000 million people will live on the earth in the year 2000 (according to the calculations of the Food and Agriculture Organization of the United Nations there will be 6500 at this time); this number would be doubled again in a further 50 years. The rapid development of technology during the last 150 years which paralleled the increase in the earth's population can best been seen from the increase in the speed of transport (*Bertaux*, *Blasius*, cf. *Fig. 39*). In the year 1830, with a speed of about 35 km per hour the railway was the fastest means of transport.

At the turn of the century one had reached a maximum speed of 100 km per hour with the automobile and 110 km per hour with the propeller airplane. However, in the next 50 years, the propulsion of vehicles and airplanes was continuously intensified and airplanes could be driven by jet propulsion faster than the speed of sound (about 1200 km/h) and several times this figure. Space rockets with a speed of about 40,000 km/h now circle the earth or are sent to the moon and to the planets. The speed of light would be the aim of this development. Even now it is speculated how man could propel space vehicles moved by the pressure of light still faster to distant stars.

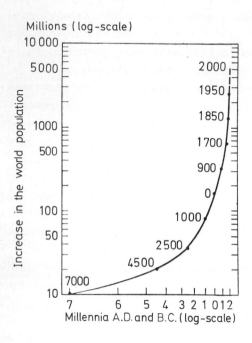

Fig. 38. Increase in the earth's population (millions of people) between 7000 B.C. and 2000 A.D. Both the earth's population and the millenia are on a log scale (modified after *Scheer*, 1958)

However, the grave consequence of this increase in acceleration of the means of transport and the huge human proliferation pressure connected with the domination of the whole of life by technology is the destruction of the previously prevailing natural equilibrium of nature, countryside, plants, animals, and humans on the earth (*Klages*,

Ziswiler). Destruction of forests, overgrazing, agricultural depletion of the soil and industrialization (to which previously fruitful regions have been sacrificed), have taken on dimensions which have never been dreamt of before. Every day in West Germany alone 300 acres of forest and field, pasture and moorland, and also parks in the cities, are destroyed in the interest of road construction, building industrial works and other technical installations (*Bruns*). The countryside often resembles a battlefield. The rivers have become sewers. Drinking-water becomes ever more scarce and expensive. Poisonous fumes from vehicles of all kinds, smoke, dust, and industrial gases pollute the air, not to mention radioactive contamination of water, soil, plants, animals and foodstuffs. One could make a long list of damage and destruction which technology and industrialization have caused.

An example which concerns everyone but particularly affect the biologists is the extinction of entire species of animals from the earth. This constitutes an irreparable loss for the whole of mankind. *Figure 40* shows the increasing tendency of destruction alone among mammals for the last 150 years (*Scheer, Ziswiler*). About 110 species from the diversity of living nature have been irreparably wiped out as a result of destruction by men, by excessive hunting, by pesticides, and so forth.

The destruction of nature together with the enormous increase in population creates situations which are threatening to become catastrophic in many countries. In India the population has grown from 258 millions in 1913 to 511 millions in 1967. However, the harvest of rice, the main food in the country, has decreased in the same period from 38 million tons per year to 29 million tons. The rice harvest per capita of the population has thus fallen from 147 kg to 66 kg or less. Famines in many parts of India demonstrate this catastrophic situation. Similar situations prevail in other developing countries.

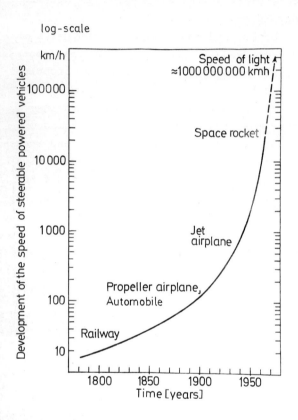

Fig. 39. Increase in the speed of steerable powered vehicles in the last 200 years of human history (after *Blasius*, 1966)

There are reputable calculations according to which even today the amount of foodstuffs which could be produced with the greatest possible utilization of the whole surface of the earth is not sufficient to feed the whole population of the earth on a European North American maintenance level. According to data of the United Nations

Organization for Food and Agriculture (FAO) about half of
the world population, i.e. about 1500 million people, suffer
from undernutrition or an incorrect composition of
their diet. This situation will rapidly deteriorate, particularly
in the developing countries. Since the world
population up to the year 2000 will grow to about 6500
million people, food production must be doubled by 1980
and trebled by the turn of the century (*Sen*).

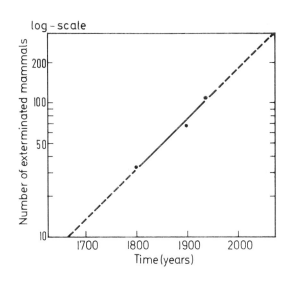

Fig. 40. Number of exterminated mammal species in the last 150 years.
1800: 33 species; 1900: 66 species; 1944: 106 species; 600 species
are almost extinct or are in grave danger (after *Scheer*, 1958). Continuous
curve: average values; broken line: extrapolation (after
Blasius, 1966)

Natural scientific study of the world has, however, not
only led to increasing technology and industrialization
which offer man many amenities and improvements to his
way of life. It has also strengthened and intensified
the means of destruction for use in war to an extent which
was previously unknown.

Mathematical representation of the number of people killed
in wars, e.g. those of Germany in the last hundred years
(*Fig. 41*), shows that a cogent law is operating here. Representations in a double logarithmic coordinate system
gives a steep straight line if one plots the war dead in
the wars of 1866, 1870-1871, 1914-1918 and 1939-1945 agains
the time in years. Since the Austro-Prussian war of 1866
(1900 dead) the number of war losses in the last war grew
to 4.2 million persons; in particular, the high proportion of civilians (about 600,000 people) show the horror
which is usual in modern wars. The crimes in the concentration camps are in a class of their own; these losses
are not contained in the figures shown. If Germany were
involved in a "traditional" war before the year 2000, the
human losses will certainly not be less than 10 millions.
In an all-out attack on our country with atomic weapons,
it is to be expected that very few people would survive.

However, one cannot conclude that the destruction of human
life on this scale is only associated with times of war,
i.e. exceptional situations, at least for the present.
Indeed the destruction of human life in times of peace
has taken on an extent which was previously unimaginable.
This can be demonstrated by the enormous increase in the
number of traffic deaths in all the industrialized countries. The statistical data on the annual total traffic
dead in West Germany since the end of the war (*Fig. 42*)
serve as an example. In a semilogarithmic representation
the number of traffic deaths increases regularly from a
minimum value in the year 1947 in a steep line to about
20,000 in the year 1971.

If one compares these annual bloody losses on the streets
with the losses in war (Fig. 41), then one makes the sad
observation that for over 25 years West Germany has been
involved in an uninterrupted war in which the amount of

losses in 3 years were the same as those of the entire
war of 1870/71. The total number of traffic deaths from
1950 to 1971 is more than 265,000, i.e. after the end of
the Second World War the population of a whole city (the
present size of Karlsruhe or Wiesbaden) has been wiped
out in traffic accidents. If one assumes a regular increase in the total number of losses, traffic deaths will
reach 24,000 per annum in the year 2000. The total number
of traffic deaths since 1950 would have risen to 860,000,
a number corresponding to the present population of Co-
logne, the fourth-largest metropolis in the Federal Re-
public or about half of the war deaths on the German side
in the First World War.

These numbers speak a dreadful and rousing language. Can
this chasm facing mankind be avoided? Can the avalanche
of destruction and death be stopped? Or will perhaps a
miracle happen, and men recall their essential tasks within nature as a whole? It is certainly wrong to believe
that mankind could renounce thinking in numbers, technology and all their splendid accomplishments on the one
hand but also their dubious and destructive results on
the other. An enormous catastrophe of famine and war would
be the certain consequence.

It will be crucial whether the rational tendencies of men
which have given rise to the "technical progress" of mankind and which is further accelerating will bring itself
into a balanced relationship with the necessities of the
whole of life on the earth; this would result in benefit
for all.

This will not be accomplished by some kind of measures
which would only accelerate the impending catastrophe,
but only through the insight of the individual human beings themselves who recognize this task and who think

and act in their life with respect for everything living
and for all of the values created by man.

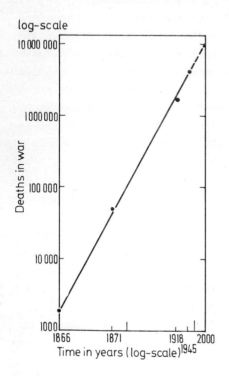

Fig. 41. German population losses through war dead in the years
1866, 1870/71, 1914-1918 and 1939-1945. Continuous line: average
values; broken line: extrapolation (after *Blasius*, 1966)

How the relations for such a "mankind in equilibrium"
would appear is perhaps clarified by the following re-
flections which I recently gave and would like to repeat
here with some additions.

Let us consider again the increase in the world population in the last century and its prospective development

in the future (*Fig. 43*). If we insert in this semilogarithmic representation the increase of participants in the international congresses of physiologists from 1889-1968 (*Blasius, Fenn*), then one sees clearly that the annual increase in number of physiologists is substantially greater than that of the world population. If this tendency is maintained the astonishing result is obtained that eventually - about the year 2350 - all human beings would be physiologists and, what is perhaps more important, all physiologists would be human beings. The penultimate conclusion may only be considered a joke, since one can assume that the physicists on the earth have a far higher annual growth rate than the physiologists! The growth of the "Landolt-Börnstein" handbook (cf. Fig. 36) is very illuminating. Hence, before the physiologists became people, an identification of physicists with the whole of mankind would have taken place! And the physiologists would be happy to be able to allowed to participate in the physicists' bandwagon of "identification with mankind".

Despite the facetiousness in which they were couched there are a few grains of bitter truth in these calculations. The growing number of people on the earth will without doubt make their conditions of life more and more difficult. Hence, one cannot rule out that in the near future almost every human being will have to acquaint himself with the laws of our earth and those of the neighboring stars. And it will also be necessary to occupy himself intensively with the facts of the human body and its environment. The survival of mankind will be largely determined by all these factors.

Of course we may assume that the increase in human beings - also of physicists and physiologists - will not proceed continuously with the same increase factor. All growth curves in the organismic world finally give way to a section with lower growth rate after an exponential increase with a large rate of growth (*v. Bertalanffy*). If this were not so, there would, for example, be more physiologists than men on the earth after the intersection of the two curves (cf. Fig. 43). This is undeniably a paradoxical situation!

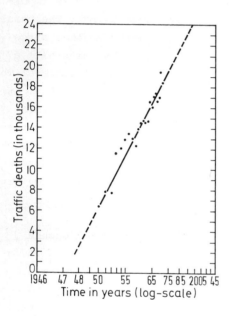

Fig. 42. Increase in the number of traffic deaths in West Germany since the Second World War. The figures were taken from publications of the Federal Office for Statistics for 1953-1971; the figures for 1950-1953 are from semi-official sources. Continuous line: average values; broken line: extrapolation (supplemented after *Blasius*, 1966)

Of course, a further accelerated growth is to be concluded from the previous growth tendency of the world population. However, in the highly developed countries it has been possible to observe in recent decades that there is a tendency for their growth trend to approach a saturation value, i.e. their population growth approaches an equilibrium.

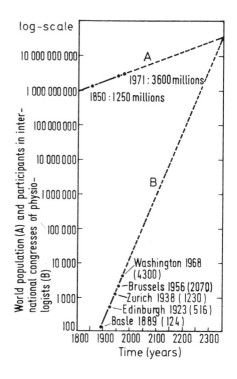

Fig. 43. Increase in the earth's population (A) (cf. Fig. 38) and increase of the number of participants in the International Congresses of Physiologists (B) in 1889, 1923, 1938, 1956 and 1968. Population and participant figure on a log scale, years on a linear scale. The fictive intersection of the extrapolated growth functions would be in the year 2350 at 35000 million inhabitants on the earth (supplemented after *Blasius*, 1966)

In order to understand this fact we must try to understand the growth curve in a closed population and the influences which determine it. The growth curve is known

to result from the difference between the birth and death
rate and the consequent rate of increase. The enormous
increase in the earth's population in the last hundred
years is to be attributed above all to an increase in the
birth rate and a fall in the death rate. The reduction
in child mortality and the increase in the average length
of life were crucial factors.

In the highly developed countries a decrease in the birth
rate can be detected in recent decades. This results in
a decrease in the growth rate with a constant low death
rate. A growth curve results which first goes through a
very steep phase after an initial slow increase and then
tails off into a smaller and smaller growth rate. This
from of growth curve for a population with time is termed
a saturation function; it is also found in other growth
processes in nature.

That the world population is now further increasing and
will continue to do so for the forseeable future is brought
about by the fact that the underdeveloped nations which
constitute a large part of the world population have a
very high growth rate. At the moment the decrease in the
growth rate in the highly developed countries has only
a slight effect on the whole.

If one investigates the increase in population growth in
a few civilized countries, then it is clear that the increase in these countries has undergone a characteristic
change in the last 150 years. In *Fig. 44*, the growth in
population of the United States of America (USA) between
1790 and 1968 is shown on a logarithmic scale. One sees
readily that the curve shows a kink about the year 1820
after an initial straight course and then flattens off
again after a slight increase about the year 1920. Certain political and economic events (Napoleonic wars, the

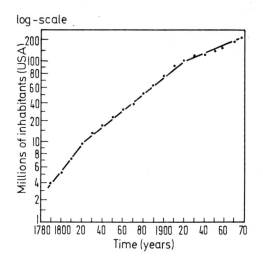

Fig. 44. Population development in the USA from 1780 to 1968. Points represent the number of inhabitants (millions) on a log scale.
—— = linear approximations (supplemented according to *Blasius*, 1970)

First World War) were clearly the important factors in this development.

The curve of population increase in the area of West Germany shows a similar course if one includes the population numbers for the same area in the former German Reich (*Fig. 45*). Distinct kinks are found in the curve about the years 1914 and 1945. The last section is of course increasingly kinked due to the major emigration from the evacuated areas in the East after the last war [1]. The population shows a distinct tendency to decrease in the area of East Germany.

---

[1] These calculations and considerations are continued in the paper by *Blasius* and *Höber*: "Mathematische Behandlung des Wachstums der Bundesrepublik Deutschland nach Ihrer Hochschulen". Deutsche Universitätszeitung, 8, 690-693 (1975); (Mathematical Analysis of Population in West Germany and Its Universities).

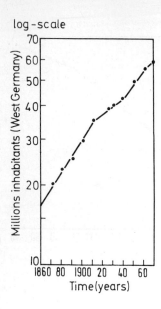

Fig. 45. Population development in West Germany from 1945 to 1965 supplemented by data for this part of Germany from 1871 to 1944. Points represent the number of inhabitants in millions in a log scale. ——— = linear approximations (assembled by *Blasius*)

Similar population curves (with almost the same increase coefficients after the last World War as for the USA) are found for the United Kingdom, France, and also the Soviet Union. Even Sweden, which has been neutral for a long time, shows a kink in the coefficient of population increase about the year 1880. Afterwards the curve becomes flatter and has an almost regular increase like the other civilized countries.

We can conclude from these calculations that with increasing civilization a population state finally results which corresponds to the possible dynamic equilibrium (*von Bertalanffy*) of continuous gain and loss. As indicated by the ever-increasing trend this state is clearly not yet reached for the total population of the earth. The proportion of inhabitants of highly industrialized countries in this total number is still too small for a state of equilibrium to be obtained.

Hence, we do not need to fear the spreading of civilization on the earth, since precisely this may constitute a reason for hope. The increase in education and the improvement of conditions of life in the underdeveloped countries will have a very decisive significance for stabilizing population trends in the earth.

The question as to whether mankind can be adequately fed in the future is optimistically assessed by the economist *Baade*. He believes that the large food deficits which have already been mentioned could be eliminated by opening up deserts, by irrigation schemes and increasing yields by means of artificial fertilizer, by improving seed and also by advances in cattle breeding. However, the precondition for such a development is that a world war is prevented, since this would result in a substantial deterioration of the situation of all mankind in every respect. *Baade*'s publications show that he estimates the maximum world population with utilization of all energy and food reserves on the earth at thirty to thirty-five thousand million people. I think that this would be a theoretical figure for the end point of the development if man comes to terms with his population growth rationally. Without doubt, the maximum world population will reach a far lower value.

If the tolerable life on earth is to be guaranteed in the future, mankind must spare no trouble in overcoming the three scourges: hunger, poverty, and war.

It is just as important that man should be in harmony with his position within the whole of living nature.

Man is distinguished from the animals by his having a spirit with which he is able to penetrate nature. As a living animate being, however, he stands in a polar living connection with all the rest of the beings in the whole world - as do these with one another; the living world is the primeval basis from which he derives all his powers. Man's special position in the natural domain as a being with spirit places responsibility on him for his thinking and actions. This means that the mental penetration of life can only be meaningful for him if it takes place in harmony with the necessities of the rest of the living domain. In view of the danger that all life on earth and all the creative works of mankind will be completely destroyed, man should no longer contribute to anything which may unleash his extermination. The dignity of man can only be maintained by a respectful attitude to the living cosmos which at the same time expresses his thankfulness for the life bestowed on him.

VII. The Essence of Health and Its Maintenance

> Healthy thinking is the greatest
> accomplishment. Wisdom consists of
> speaking the truth and acting in
> accordance with nature - listening
> to her.
>
> Heraclitus

The word "health" has never been so overused as in our time; I am thinking for example of the innumerable advertisements in the newspapers, on the radio and television for foods, medications and equipment which are supposed to make people "healthy again very easily". At no time have the efforts concerned with human health been so great, i.e. so large-scale and comprehensive as at present. I draw attention to the increasing number of doctors who are concerned about curing a greater number of their patients, and also of the efforts of governments to build ever larger hospitals and preventive facilities. A total registration of the health of everyone is planned and being put more and more into practice.

We might well conclude from all these efforts that human health is in a bad way. Are the intensity of preventive medicine and concern about health perhaps caused by the fact that nobody really knows what health actually is? It almost appears so. It is characteristic and highly illuminating that in the sciences which are concerned with life, i.e. "healthy" life, the concept of "health" does not occur at all. In the relevant textbooks of hygiene, disease, physiology and medicine the word "health" is not

mentioned although hygiene is derived from *Hygieia*, the
Greek goddess of health. I have closely examined about
18 modern textbooks in these disciplines in this regard.

I did not find any keyword "health" in "Knaurs Gesundheitslexikon". "Der Große Brockhaus" and "Der Gesundheits-
Brockhaus" at least state that it is difficult if not impossible to define unequivocally the commonly used word
"health", designating one of the greatest blessings of
men, and to define it as a logically unassailable concept.

The 23-volume "Encyclopaedia Britannica" defines the concept "health" in a single line: "Health is a condition
of physical and mental well-being". As we shall see this
is a highly unsatisfactory definition of health. "Well-
being" indicates situations which give rise to precisely
the opposite of health, as we know very well today.

What is the reason why it is not possible to give an exact
definition of health? The answer must be: because health
is the same meaning as life and life cannot be precisely
defined either.

However, if we cannot define health precisely, then it
will perhaps be possible to describe it and to interpret
it according to its meaning.

Linguistics usually gives valuable information on how
to find the meaning or sense of a word or to recover them.
The new High-German word "gesund" is related to the word
"geschwind" which in Old High German and even today has
the meaning "moved", "powerful", "rapid". According to
its meaning, health could consequently be viewed and interpreted as "powerful mobility". We found the same statement with *Alkmaion*: Healthy life is "moved of itself"
(cf. Chapter III). If health is "spontaneous movement"

as we observed, the loss of the faculty of movement is to be viewed as disease. The most important task of the doctor would thus consist in restoring living movement.

It was emphasized in Chapters II and III that living movement is to be viewed as rhythmic movement. Rhythm is "renewal of the similar at similar time intervals" not a "repetition of the same at the same time intervals"; rhythm is thus to be separated from period or tact (*Klages*).

In order to understand the origin of rhythm or rhythmic movement, it is necessary to explain the polarity of all living events. The rhythm of life in its innermost essence is expressed in this polarity.

The fundamentals of the interpretation of life as a continuum of living poles can be demonstrated, even with the arguments from Greek natural philosophers *Heraclitus* and *Protagoras*. In the late Middle Ages and in the Renaissance *Paracelsus* and other physicians and philosophers resumed thinking in polarities. But only *Goethe* and the romantics *Novalis*, *Carus* and *Bachofen* were able really to bring this natural philosophical interpretation of life (which is in opposition to the rationalistic theory of life) to a true breakthrough. The legacy of romanticism was carried further and perfected by *Nietzsche* and *Klages* in their philosophies of life.

These philosophers and particularly *Klages* start from the trinity of human life as it was already laid at the basis of interpretation of life by the Greek natural philosophers; they distinguish σωμα, the body, ψυχη, the soul, and νους, the spirit. Only the soul is characterized by the faculty of contemplation, in that it brings the effective impressions of the world of appearances into fusion with the images available in the soul. In this fusion the essence of the polarity of experiencing soul and acting image becomes entirely clear.

The polarity also shows itself in the estranging of image and soul that can again follow. In the contemplative state, however, the estranging and fusion of soul and image stand in equilibrium (*Klages*).

Both the fusion and the incompleteness of the fusion prove that an attraction goes out from the image which acts on the embodied soul, causing it to carry out movements with the aim of touching or entirely incorporating the embodied image.

Consequently the essence of polarity can be interpreted as elementary *Eros* (*Klages*), as he already appeared to the Greeks in their cosmogony as the third divine power at the beginning of the world apart from the first deities: *Chaos* (the unformed) and the goddess *Gaia* (the formed earth) (*Blasius*, 1966).

While the soul only receives influences from the embodied image, i.e. from the appearance of the image, on the other hand νοῦς, the spirit alone depends directly on influences of the soul, i.e. it is only through these influences that it becomes aware of what the soul has sensed from reality itself. However, apart from the influences of images the soul can also experience impulses from the spirit.

The soul is consequently through and through receptive, i.e. the center of gravity of its essence is in what is sensed: the soul echoes from sensed images. On the other hand, the essence of the spirit consists above all in action, i.e. it becomes the faculty for action by coupling with the living soul. The activity of the soul does not take place arbitrarily but from necessity; it is a must. On the other hand, spirit activity is sometimes expressed as an independent and sometimes as an extorted action (*Klages*).

In the connection of the poles of influencing world and receiving soul, which determine one another, we find the sense of experience and the not further solvable meaning of all "sense" generally. "Meaningful" is the connection of the distinguishable members of a composite; and the most original composite is ambivalence. The return to this is promoted by the polarity of the appearing reality and the soul which experiences it; a polarity without which there would be no interpretative description of living essence and of the thinking being.

Living connections are hence recognized in manifold, natural polarities. Thus, the male being is polarly related to the female being, the body to the soul, the fusion to the estranging, contemplation to action, pleasure to pain, genesis to passing away and much else (cf. Chapter I, p. 6).

However, the soul does not have the same relationship to the body as cause to action, but that of the meaning to the appearance of meaning (*Klages*). Neither the soul nor the body act on each other since neither belongs to the world of thought things. If the sequence of cause and effect merely signifies a thought or illusory relationship between parts of a continuum which was separated in thoughts, then meaning and the appearance of meaning are themselves a connection.

Further examples of such polarities were already mentioned in Chapter I, where it was indicated that the axioms of physics are also found among these polarly divergent tendencies of life. They cannot be further explained by natural science, i.e. logically and causally and must hence be accepted as given. The idea of polarity leads of necessity to the concept that the whole cosmos is living.

We have thus far interpreted and described the essence of health as "powerful movedness", i.e. as "rhythmic movedness", which always occurs between two coordinate poles. It has further become understandable to us that the essence of health is above all the rhythmic interplay between the appearing reality and the sensing soul.

If we wish to discuss the living, powerful rhythmic movements of man, then this can be in reference to his body and his soul on the one hand and also to his spirit on the other. We could therefore say that the living, rhythmic movement of all three domains of the human essence constitutes health.

The living movement of the body can consequently never represent the essence of health itself, because the movements of the body are always intimately connected with those of the soul and the spirit. There are thus no movements of the living body which do not express simultaneously a psychic or spiritual movement of the human being. Both excessive movement and lack of physical movement have their psychic and spiritual expression. These relationships constitute the basis of inferring the intensity, the cause, and the duration of an emotion from an expressive movement, because each bodily movement realizes the Gestalt of such an emotion. These stirrings of the soul always include the images and characters of the world as polar equivalents. Life can only be understood as an experience of living reality. Without this experiencing of living reality it is also not possible to obtain an appro-

priate concept of reality. Experience is thus always a
precondition for conceptual thinking. We can also say that
without experience thinking is not possible; life and experience are the necessary preconditions of all conceptual
thinking.

If the proper movement of a living being is to be seen
as a meaningful characteristic of its vitality and health,
this principle also allows the development of diverse locations for the origin of movement and its meaning for
human beings. The linguistic expressions which describe
the movements of the body, the soul and the spirit give
rise to important conclusions:

The body expresses *movement*, the soul *movedness* (emotion)
and the spirit *mobility*. One says the soul is "moved",
but the spirit is "mobile". We refer to the "heart's being moved", not to the "mobile heart". The word "soul"
is derived from the word αἴολος = moved (*Aiolos* is the
name of one of the wind gods).

The character of the mobility of the spirit is expressed
linguistically in very different ways (*Hönel*); as dexterity, nimbleness, flexibility, skill or alertness, "buisiness", and in a derogatory sense as wheeling and dealing
or vileness; with a strong component of will as enterprise
or presence of spirit. In contrast to this is immobility
of spirit which is coupled with strong willpower and characterized as steadfastness, perseverance, imperturbability, pertinacity, doggedness, strength of character. If
the immobility has reached its highest degree and thus
leads to inhibition of feelings, the characteristics are
designated intractability, stubbornness, pigheadedness,
moroseness, and inflexibility.

A series of ways in which the soul and feelings are moved in a similar way are established: e.g. emotional lability, arousal, passionate happiness, and with emphasis on excess, towering rage, lack of restraint and hedonistic behavior.

The lack of emotions on the other hand, is designated as lack of feelings, of excitability, of happiness, or of enthusiasm.

After these characterizations of the movedness of the soul and the mobility of the spirit we wish to turn to the movements of the body: movements of expression, voluntary movements and reflex movements (*Klages*). Only movements of expression have the full vital content because they correspond quite polarly to the psychic impulses. In voluntary movements the mental, i.e. voluntary component, is so prominent that the psychic impulses are inhibited. One thinks of the movements which are learned in military training. The reflex movements which under natural conditions are usually incorporated into or subordinated to the movements of expression and voluntary movements, remain purely physical if they are isolated or induced experimentally (*Blasius*). They hence have a mechanical, automatic and predictable character. The inducing stimulus and the physical excitation which takes place occur in the individual organs.

Movements of expression on the contrary are expressions of the whole living being and hence may only be related to it. When they are particularly vital they do not show those characteristics which are specific to reflex movements (*Blasius*).

Hence, the more living a movement appears, the less it can be calculated and predicted. The expressive movements must thus be characterized and evaluated according to the sense and meaning of their psychic and mental content, according to their rhythm and their polar relationships.

It would certainly be worthwile to describe the movements of individual organs, disorders of their movement and the partial or total loss of movement. Further overshoot and excessive movements would have to be described in this context.

From the large number of disorders of bodily movements, however, I will only select those which appear to be particularly important for the assessment of human health in our time.

The striking lack of physical movement not only in adults but also in young people and even in children must be mentioned first: this has led to a serious reduction of physical efficiency and the faculty for psychic expression. Both in largely sedentary activity and in forms of work with automatic machines or monotonous work processes there is loss of bodily movement because only individual muscle groups are active and a harmonic balance with alternating activation of all the muscles of the body is entirely lacking. In addition, exercise and promotion of the organs of circulation and respiration, the glands of the body and the central nervous system are also neglected in such work forms, so that the efficiency of the body as a whole substantially diminishes. Purely passive means of transport such as the car, bus or railway also have disadvantageous results for the whole organism, because travel with these comfortable means of transport obviates the body's own movements. After a working day with little movement (the expression "working" day is often no longer appropriate!) the family assembles before the television screen and once again only motionless watches; there can only be unfavorable consequences of such a way of life. Physiologists and physicians hence observe an excessive increase in body weight with accompanying disturbances of circulation, metabolic anomalies, disorders of digestion, foot trouble and other bodily and psychic disease sequelae of inadequate physical activity. I must be satisfied here with the few indications, although these injuries are the most serious for mankind today.

On the basis of long years of experience, *H. Mommsen* showed that the natural bacteria in the mouth and intestines of man are related to a further important disturbance of health - i.e. "powerful movedness". We can view this disturbance of health as a disturbance of the polar connection of man to the living creatures naturally occurring in his body. As is known since the discovery of the cell bacteria by *Escherich* (1886) these bacteria play a very important role in the degradation in the intestine of certain substances taken up by man with the food. These bacteria are indeed necessary for the life of the human body since they form a vitamin which the body must receive from them because it cannot synthesize it itself. This is vitamin K, which is involved in the formation of prothrombin (this is in turn important for blood coagulation).

As could be demonstrated, a healthy bacterial flora in the mouth is also vitally necessary for man. *Mommsen* observed that the destruction of this oral flora by administration of antibiotics had a very unfavorable effect on the health of children; an effect which could only be removed by the restoration of a healthy mouth flora.

Such experiences can be compared to analogous observations on viruses which include both the excitants of certain diseases and also germs which are very important for normal body metabolism, i.e. for its health (cf. Chapter I, p. 29). Such viruses do not have a cellular structure: rather they consist of particular molecules or molecular aggregates. Although viruses do not have a cellular structure they are capable of proliferation. Outside a cell they do not have this faculty, in contrast to bacteria which show a distinct tendency to proliferate even outside cells.

If viruses get inside a cell body they suddenly proliferate very intensively: the substances are supplied by the host cell and the virus is only the matrix which stamps the new virus with the image of the old. This is one of the many wonderful ways in which life forms new life and everpresent polarity becomes effective!

A third example of the polarity of germs and infected organism which is important for the recovery of health is a fever which is known to be elicited by penetration of germs into the body. It sets in motion a considerable movement of the cardiovascular system, of metabolism, of the heat economy and the nervous system. Very experienced physicians consider that fever kept in moderate limits introduces a healing process which should not be interrupted except in emergency. By the schematic use of antibiotics this natural healing process is often interrupted and the healing action of the fever (which is accompanied by a formation of certain defence substances and by an activation of the body) comes to a stop. The bacteria in the body are destroyed, but the natural healing process is not yet complete. Relapses are thus a frequent result of interrupting the fever and the recovery of the patient remains incomplete (*Mommsen*).

If only a few possibilities for maintaining human health have become clear from the examples of lack of bodily movement, maintenance of natural symbionts and the healing effect of fever, I would now like to discuss the three domains of "powerful movedness" in reference to man, which are the most important preconditions for the maintenance of his body, his soul and his spirit.

To maintain life, proper vital movement of the living creature and a rhythmic polar connection to all essential domains of his being are necessary. These domains correspond to the body, to the soul and to the spirit.

Let us begin with the equivalence of the body. The body is maintained by a meaningful adapted, natural diet. Physiologists and physicians would have a great deal to say on this subject, since this is the area in which mistakes are frequently made.

One cannot emphasize enough the health-promoting effect of physical activity in free nature, since the changing features of the countryside, the weather, the climate, light and water constitute the best rhythmic stimulators for the rhythmic movements of the body. Work and movement in fresh air are hence superior to all other forms of work and movement in their strengthening effect on the body. This applies in particular to play of children, to rambling, mountain climbing, swimming, riding, rowing and all kinds of sport and play which can be carried out in the open air.

Every physical activity should always be preceded by an adequate period of sleep and rest, since full efficiency requires complete recovery. In order to increase physical performances a tiring muscular activity is a precondition: without fatigue there can be no increase and adaptation of the musculature and also no healthy sleep. This topic was thoroughly discussed in Chapter III. We will hence restrict ourselves to these indications.

We shall now inquire into the conditions of life which enable the soul of man to gain and retain its movedness, i.e. its health. These conditions of life can only be evaluated according to the equivalences which enable them to supply polarity to the soul.

Klages has stated with good reason that such an evaluation must incorporate everything which serves for the welfare of life. However, no psychagog (Seelenführer) should imagine that he can change or improve the essence of a human being. From a fir-cone grows a fir-tree, from an acorn an oak; the person who cares for the seed is never the creator of the form or the producer of the growth. A plant needs light and moisture, and it becomes more or less splendid depending on how we attend to both. Guidance of the soul hence consists of supplying psychic food.

Klages enumerates the main foods of the soul: wonder, love and example. The soul finds wonder in landscapes, in poetry, in beauty. In all three domains the living images viewed by the soul act as a wonder. In Germany the word for beauty (Schönheit) expresses this truth very well, since "Schönheit" is derived from "schauen" and means roughly: "geschaute Seele" (contemplative soul). The soul is exposed to scenery, poetry, and beauty and one sees whether it flourishes from them.

And love in the widest sense of the word (which covers estimation, admiration, indeed every kind of affectionate regard) is truly effective only from loving persons. The eternal image of such guidance of the soul is the picture of the loving mother with the beloved child. One thus gives the soul all the irradiation of motherly love and sees whether it blossoms.

Klages accounts the gods, the poets and heroes as exemplary models. He had certainly thought of those gods who appear incarnate to men and which have given happiness by their being, by their health and their gifts. As heroes he certainly meant both those who deliver from need and also great creative spirits, musicians such as *Bach*, *Mozart* and *Beethoven*, sculptors and researchers such as *Carus* and *Bachofen* and leaders of the soul such as *Goethe* and *Nietzsche*.

The soul is allowed to have a view of the heroes, continues *Klages*, and one sees whether it blossoms. And if it does not blossom anywhere then it does not have any blossoming force - no leaders of the soul can help such souls.

"This is the secret of the soul; it only becomes richer in giving. It is not the love which it receives but the love which is sparked in it by received love. This is what feeds the soul. And all the wonders and examples of the world remain only a theatrical play, if they are not able to awake in the soul the secret wonder and the secret hero".

These words have - I believe - shown sufficiently what importance the movedness of the soul, the emotions and the heart has for true human health. We still should say something about the importance to health of the mobility of the spirit. One must distinguish in the spirit the cognitive capacity and the will. The cognitive capacity serves the conceptual apprehension of the world and the will serves the active apprehension of the world. The mobility of action is pronounced by the doers; that of thinking by the thinkers. *Goethe* observes: "Action is easy, thinking is difficult". The apprehension of the world by the will is so easy for many people that the thought which should precede the action is neglected. On the other hand there are many thinkers who devote their time only to their thoughts and forget or indeed are afraid of action.

That mental movedness would be harmonious which could include the both domains of the mental and at the same time feel itself bound to life. Mental effort and activity in the service of life are indeed the life style which should be put into practice for people in our time to maintain their health and the health of the whole nature!

If we remember the sad fact to what extent the life of the plants, animals, humans, the scenery, water and the air nowadays are curtailed, eradicated, destroyed, poisoned and infected, then we can expect a decisive help only from the renewal of human souls which can be awakened alone by the polarity with the living reality, i.e. by the love of life. The possible overstraining of the performances of human will allows the supposition that men could one day succeed in eradicating all plant, animal and human life on the earth (cf. Chap. VI).

To find out what life and health essentially means we required a great deal of thought. With the knowledge thus obtained we gain a yet more vital impression from the lyrical words of the young *Goethe* from the year 1782 in which he describes what he meant by nature, which he equated with life [1]:

"Nature! We are surrounded and embraced by her - unable to emerge from her and unable to penetrate more deeply into her. Unasked and unheralded she takes us up in the circle of her dance and pushes forward with us until we

---

[1] As an old man *Goethe* could not say whether this hymn-like essay which is incorporated in the handwritten "Tierfurter Journal" (of which the most important parts are reproduced here) was his own work. It appears that Georg Christoph *Tobler* from Zurich (1757-1812) had written it after conversations with him and *Goethe* then edited the pages for the journal.

are tired and fall from her arms. She is forever creating
new forms; what she now is was never there before; what was
will never come again - everything is new and yet always
old.

We live in her midst and are foreign to her. She speaks
incessantly to us and does not betray her secret. We act
constantly on her and yet have no power over her.

She seems to have founded everything on individuality
and does not make anything out of the individuals. She
is always building and always destroying; her workshop
is inaccessible.

Every gift she makes is out of kindness; she only gives
what is indispensable. She lingers so that she is desired;
she hastens, so that we are never tired of her.

She has neither language nor speech; but she creates tongues and hearts through which she feels and speaks.

Her crown is love. Only through love does one approach
her. She makes abysses between all beings and intertwines
everything. She has isolated everything and draws everything together. With a few sips from the beaker of love
she compensates for a life full of pain.

She is everything. She rewards herself and punishes herself, she enjoys herself and tortures herself. She is
rough and gentle, sweet and dreadful, powerless and allpowerful. Everything is always present in her. She does
not know past and future. The presence is her eternity.
She is benevolent. I glorify her and all her work. She
is wise and still. One cannot extract any explanation from
her or any gift that she does not freely give. She is
artful, but to good purpose, and it is best not to notice
her cunning.

She is entire and yet never completed. As she proceeds, so can she always proceed.

She appears to everyone in a special form. She hides in a thousand names and terms and is always the same.

She put me here and she will also lead me away. I trust in her. She can do with me what she will. She will not hate her work. I was not speaking of her. No, she has spoken everything that is true and false. Everything is her failing, everything is her merit".

VIII. Bibliography

I. Epistemological and Methodological Foundations of Research into Life

BERTALANFFY, L.v.: Theoretische Biologie, Vol. II, 2nd ed. Berlin 1951

BLASIUS, W.: Erkenntnistheoretische und methodologische Grundlagen der Physiologie. In: LANDOIS-ROSEMANN: Lehrbuch der Physiologie des Menschen. Vol. II, 28th ed., pp. 990-1011. München-Berlin 1962

BLASIUS, W.: Die Klages'sche Lebenslehre als Ergänzung der naturwissenschaftlichen Lehre vom Leben. Zschr. Menschenkunde 27, 193-204 (1963)

BLASIUS, W.: Zur Geschichte der Reflexlehre unter besonderer Würdigung des Beitrages von Paul Hoffmann. Dtsch. Zschr. Nervenhk., 186, 475-495 (1965)

BLASIUS, W.: Über die Grenzen der naturwissenschaftlichen und der naturphilosophischen Lehre vom Leben. Schweizer. Ärztezeitung 49, 1125-1132 (1965)

BLASIUS, W.: A propos de la pensée inspirée par les sciences physiques et naturelles et inspirée par la philosophie de la nature dans la science de la vie. Revue de Médecine Fonctionnelle, pp. 31-53 (Paris) 1969

ELSASSER, W.M.: The physical Foundation of Biology. 1958

FRAUCHIGER, E.: Die Bedeutung der Seelenkunde von Klages für Biologie und Medizin. Bern 1947

KLAGES, L.: Vom Wesen des Bewußtseins. Aus einer lebenswissenschaftlichen Vorlesung. 4th ed. München 1955

KRETSCHMER, E.: Körperbau und Charakter. 25th ed. Berlin-Göttingen-Heidelberg 1967

NEUMANN, J.v.: Cerebral Mechanics in Behavior. The Hixon Symposium. New York 1951

NEUMANN, J.v.: The Computer and the Brain. New Haven 1958. Translation München 1960

PALÁGYI, M.: Naturphilosophische Vorlesungen über die Grundprobleme des Bewußtseins und des Lebens. 1st ed. 1907, 2nd ed. Leipzig 1924

PRIGOGINE, J.: Etude Thermodynamique des Phénomènes irréversibles. Paris and Liège 1947

RASHEVSKY, N.: Mathematical Biophysics. 2nd ed. Chicago 1948

REPGES, R.: Grenzen einer informationstheoretischen Interpretation des Organismus. Gießener Hochschulblätter 6, Vol. 3/4 (1962)

ROTHSCHUH, K.E.: Geschichte der Physiologie. Berlin-Göttingen-Heidelberg 1953
SCHRÖDINGER, E.: Was ist Leben? Bern 1946
WIENER, N.: Mensch und Menschmaschine. Original: The Human Use of Human Beings 1958
WIESER, W.: Organismen, Strukturen, Maschinen. Frankfurt 1959

II. *Rhythm and Polarity - Physiological Analysis and Phenomenological Interpretation*

ASCHOFF, J.: Gesetzmäßigkeiten der biologischen Tagesperiodik. Dtsch. Med. Wschr. 88, 1930-1937 (1963)
ASCHOFF, J.: Vortrag bei der 32. Tagung der Deutschen Physiologischen Gesellschaft vom 4.-8.10.1966 in Berlin
ASCHOFF, J., DIEHL, I., GERECKE, U., WEVER, R.: Aktivitätsperiodik von Buchfinken (Pringilla coelebs L.) unter konstanten Bedingungen. Z. vergl. Physiol. 45, 605-617 (1962)
ASCHOFF, J., WEVER, R.: Beginn und Ende der täglichen Aktivität freilebender Vögel. J. Ornithol. 103, 2 (1962)
ASCHOFF, J., WEVER, R.: Spontanperiodik des Menschen bei Ausschluß aller Zeitgeber. Naturwissenschaften 49, 337-342 (1962)
BLASIUS, W.: Erkenntnistheoretische und methodologische Grundlagen der Physiologie. In: LANDOIS-ROSEMANN: Lehrbuch der Physiologie des Menschen. Vol. II, 28th ed., pp. 990-1011. München-Berlin 1962
BLASIUS, W.: Ganzheiten und Systeme und ihre Grenzen. Vitalstoffe 10, 152-155 (1965)
BLASIUS, W.: Rhythmus und Polarität - physiologische Analyse und erscheinungswissenschaftliche Deutung. Dtsch. Med. Wschr. 93, 1489-1495 (1968)
BLASIUS, W., ALBERS, C., BACH, G., BRENDEL, W., THAUER, R., USINGER, W.: Größe und zeitliche Verteilung der Spannungsproduktion des Herzens während Hypothermie beim Hund. Verh. Dtsch. Ges. f. Kreislauff. 23, 135-142 (1957)
BLASIUS, W., ALBERS, C., BACH, G., BRENDEL, W., THAUER, R., USINGER, W.: On Cardiac Electrophysiology in Hypothermia. Exp. Med. a. Surg. 19, 258-269 (1961)
KLAGES, L.: Vom Wesen des Rhythmus. Kampen a. Sylt 1934
KLAGES, L.: Der Geist als Widersacher der Seele. 3rd ed. impr. ed., München and Bonn 1954
ULMER, F., KOENIG, W., BINDER, E., **HENDRIOK**, H.: Kreislauf in Hyperthermie. Pflügers Arch. ges. Physiol. 276, 66-81 (1962)
WEVER, R.: Zum Mechanismus der biologischen 24-Stunden-Periodik. Kybernetik 1, 213-231 (1963)

III. *Bodily Movement and Exercise - Physiological Analysis and Phenomenological Interpretation*

BLASIUS, W.: Rückenmark. In: LANDOIS-ROSEMANN: Lehrbuch der Physiologie des Menschen. Vol. II, 28th ed., pp. 663-694. München-Berlin 1962

BLASIUS, W.: Erkenntnistheoretische und methodologische Grundlagen der Physiologie. In: LANDOIS-ROSEMANN: Lehrbuch der Physiologie des Menschen. Vol. II, 28th ed., pp. 990-1011. München-Berlin 1962

BLASIUS, W.: Über die Grenzen der naturwissenschaftlichen und der naturphilosophischen Lehre vom Leben. Schweizer. Ärztezeitung 49, 1125-1132 (1965)

BLASIUS, W.: Über die Grenzen der naturwissenschaftlichen und der naturphilosophischen Lehre vom Leben. Dtsch. Ärztezeitung 64, 2470-2475 (1967)

BLASIUS, W.: Rhythmus und Polarität - physiologische Analyse und erscheinungswissenschaftliche Deutung. Dtsch. Med. Wschr. 93, 1489-1495 (1968)

BLASIUS, W.: Die physiologischen Grundlagen der Übungsbehandlung. Zschr. Physik. Med. 1, 394-412 (1970)

GAUER, O.H.: Kreislauf des Blutes. In: LANDOIS-ROSEMANN: Lehrbuch der Physiologie des Menschen. Vol. I, 28th ed., pp. 93 a. 95. München-Berlin 1960

GÖPFERT, H.: Die physiologischen Vorgänge bei der Muskelspannung und -entspannung. Arch. physik. Therap. 18, 249-259 (1966)

RANKE, O.F.: Arbeits- und Wehrphysiologie. Leipzig 1941

IV. *Human Language - Physiological Analysis and Phenomenological Interpretation*

BERGER, H.: Das Elektroencephalogramm des Menschen. Nova Acta Leopoldina (Halle) 6, 173 (1938)

BLASIUS, W.: Erkenntnistheoretische und methodologische Grundlagen der Physiologie. In: LANDOIS-ROSEMANN: Lehrbuch der Physiologie des Menschen. Vol. II, 28th ed., pp. 990-1011. München-Berlin 1962

BLASIUS, W.: Die Klages'sche Lebenslehre als Ergänzung der naturwissenschaftlichen Lehre vom Leben. Z. f. Menschenkunde 27, 193-204 (1963). Nachdruck in: "Hestia" 1963/64 "Vorträge über das Werk von Klages" 17-28 (1964)

BLASIUS, W.: Über die Grenzen der naturwissenschaftlichen und naturphilosophischen Lehre vom Leben. Schweizer. Ärztezeitung 49, 1125-1132 (1965)

BLASIUS, W.: Ehrfurcht vor dem Leben und Zerstörung des Lebens. Hippokrates 37, 483-489 (1966)

BLASIUS, W.: Die menschliche Sprache - physiologische Analyse und erscheinungswissenschaftliche Deutung. Z. f. Menschenkunde 31, 134-157 (1967)

CLARA, M.: Das Nervensystem des Menschen. 3rd ed. Leipzig 1959

DANDY: Hirnchirurgie, 1938 (zit. nach Clara)

GLEES, P.: Gehirn. In: LANDOIS-ROSEMANN: Lehrbuch der Physiologie des Menschen. Vol. II, 28th ed., pp. 716-760. München-Berlin 1962

GRÜTZMACHER, M., LOTTERMOSER, W.: Die Verwendung des Tonhöhenschreibers bei mathematischen, phonetischen und musikalischen Aufgaben. Akust. Z. 3, 183 (1938)

KATSUKI, I.: The function of the phonatory muscles. Jap. J. Physiol. 1, 29 (1950)

KLAGES, L.: Der Geist als Widersacher der Seele. 3rd impr. ed. München-Bonn 1954
KLAGES, L.: Vom Wesen des Bewußtseins. Aus einer lebenswissenschaftlichen Vorlesung. 4th ed. München 1955
KLAGES, L.: Die Sprache als Quell der Seelenkunde. 2nd ed. Stuttgart 1959
KLAGES, L.: Ausdruckskunde. Sämtliche Werke. Vol. 6. Bonn 1964
KUGLER, J.: Elektroencephalographie in Klinik und Praxis. 2nd ed. Stuttgart 1966
PANSE, Fr.: Untersuchungen im körperlich-seelischen Grenzbereich. Bild d. Wiss. 4, 57 (1967)
RANKE, O.F., LULLIES, H.: Gehör - Stimme - Sprache. Lehrbuch der Physiologie, ed. by W. Trendelenburg and E. Schütz. Berlin-Göttingen-Heidelberg 1953
RAUBER-KOPSCH: Lehrbuch und Atlas der Anatomie des Menschen. 19th ed., Vol. II. Stuttgart 1955
REIN, H., SCHNEIDER, M.: Einführung in die Physiologie des Menschen. Vol. 16. Berlin-Göttingen-Heidelberg 1971
ROHRACHER, H.: Einführung in die Psychologie. Wien-Innsbruck 1963
WALTER, W.G.: Das lebende Gehirn - Entwicklung und Funktion. Köln-Berlin 1961

V. Stereoscopic Vision and Color Discrimination: Their Typological Polarity and Relations to Pictorial Creativeness

ALLAN, E.C., GUILFORD, J.P.: Factors determining the affective value of colors in combination. Amer. J. Psychol. 48, 643-648 (1936)
BLASIUS, W.: Das Raumsehvermögen bei Form- und Farbbetrachtern. Z. Sinnesphysiol. 70, 52-74 (1943)
BLASIUS, W., ZIEGENHAIN, L.: Eine neue Methode zur Prüfung der Farbunterscheidungsfähigkeit. Pflügers Arch. ges. Physiol. 297, 99 (1967)
BLASIUS, W.: Raumsehvermögen und Farbunterscheidungsfähigkeit als typologische Polaritäten in ihrer Beziehung zur bildnerischen Gestaltung. Arch. f. Psychol. 122, 67-91 (1970)
BLASIUS, W.: Colour Discrimination and Stereoscopic Vision in Their Complementarity and in Their Relation to Types of Pictorial Creativeness. Tag.-Ber. Int. Farbtagung COLOR 69. Vol I, 148-155. Stockholm 1969
EYSENCK, H.J.: A critical and experimental study of color preference. Amer. J. Psychol. 54, 385-394 (1941)
GOETHE, J.W.v.: Die Farbenlehre. Weimar 1827
GUILFORD, J.P., SMITH, P.: A system of color preference. Amer. J. Psychol. 72, 487-502 (1959)
HEISS, E.: Über psychische Farbwirkungen. Studium generale 13 (1960)
HOFSTÄTTER, P.R., LÜBBERT, H.: Die Eindrucksqualitäten von Farben. Z. diagnost. Psychol. 6, 211-227 (1958)
KLAGES, L.: Vom Wesen des Bewußtseins. 4th ed. München 1955
KOCH, Eb.: Ein neues Raumsehprüfgerät. Luftfahrtmed. 5, 317-321 (1941)
KRETSCHMER, E.: Körperbau und Charakter. 25th ed. Berlin 1967
LAMPARTER, H.: Typische Formen bildhafter Gestaltung. Z. Psychol. Erg.-Bd. 22, 217-356 (1932)
OSTWALD, W.: Die Farbenlehre. Berlin-Leipzig 1923

PALÁGYI, M.: Wahrnehmungslehre. Leipzig 1925
ROLLMANN, W.: Zwei neue stereoskopische Methoden. Poggendorff Ann. 90, 186-187 (1853)
SCHOLL, R., VOLLMER, O.: In: O. Kroh: Experimentelle Beiträge zur Typenkunde. Bd. I. Z. Psychol. Erg.-Bd. 14, 295 (1929)
VELHAGEN, K.: Tafeln zur Prüfung des Farbsinnes. Stuttgart 1970
WEIZSÄCKER, V.v.: Der Gestaltkreis. 3rd ed. Stuttgart 1968
WELLEK, A.: Die Polarität im Aufbau des Charakters. 3rd ed. Bern-München 1966
ZIEGENHAIN, L., BLASIUS, W.: Farbunterscheidungsfähigkeit und Raumsehvermögen in ihrem komplementären Verhalten und ihren Beziehungen zur Art der bildnerischen Gestaltung. Pflügers Arch. ges. Physiol. 297, 87 (1967)

VI. Quantitative Thinking in the Life Research

BAADE, F.: Der Wettlauf zum Jahre 2000. Oldenburg 1960
BERTALANFFY, L.v.: Theoretische Biologie. Vol. II: Stoffwechsel, Wachstum. 2nd ed. Bern 1951
BERTALANFFY, L.v.: Wachstum. Hdb. d. Zool. Vol. VIII, Part 4, p. 1-68. Berlin 1957
BERTAUX, P.: Mutation der Menschheit - Diagnosen und Prognosen, p. 62. Frankfurt-Hamburg 1963
BLASIUS, W.: Erkenntnistheoretische und methodologische Grundlagen der Physiologie. In: LANDOIS-ROSEMANN: Lehrbuch der Physiologie des Menschen. Vol. II, 28th ed., p. 990-1011. München-Berlin 1962
BLASIUS, W.: Ehrfurcht vor dem Leben und Zerstörung des Lebens - Gedanken und Berechnungen eines Physiologen. Hippokrates 37, 483-489 (1966)
BLASIUS, W.: Das Denken in Zahlen. Bild der Wissenschaft 3, 636-643 (1966)
BLASIUS, W.: Das Denken in Zahlen und seine Folgen für die Menschheit. In: Gefährdete Schöpfung, p. 43-62. Homburg and Zürich 1970
BLASIUS, W.: Das Denken in Zahlen und seine Folgen für die Menschheit. Hess. Ärztebl. 31, 854-871 (1970)
BRUNS, H.: Naturschutz ist Lebensschutz. Das Leben. Zschr. f. Biologie u. Lebensschutz 1, 14-16 (1964)
Das Gesundheitswesen der BRD. Vol. 4. Stuttgart-Mainz. Kohlhammer 1970
Der Große Brockhaus. 16th ed. Vol. XII. Artikel "Weltkriege". Wiesbaden 1957
Données statistiques - statistical Data 1960. Section de la documentation. Strasbourg, May 1961
FENN, W.O.: History of the International Congresses of Physiological Sciences 1889-1968, p. 724. Maryland: Baltimore 1968
Handbook of Biological Data. Editor W.S. Spektor. Ohio 1956
Handbook of Circulation. Analysis and Compilation by P.L. Altman. Ohio 1960
Historisk Statistik for Sverige - Historical Statistics of Sweden. Befolkning Population 1720-1950. Statistika Centralbyrau. Stockholm 1955

Historical Statistics of the United States, Colonial Time to 1957. Prepared by the Bureau of the Census with Cooperative of the Social Science Research Council. Second Printing. Washington D.C. 1961

KLAGES, L.: Mensch und Erde. 5th ed., Stuttgart 1956

LANDOLT-BÖRNSTEIN: Zahlenwerte und Funktionen aus Physik, Chemie, Astronomie, Geophysik und Technik. 6th ed. Berlin-Göttingen-Heidelberg 1950-1964

ROTHSCHUH, K.E., SCHÄFER, A.: Quantitative Untersuchungen über die Entwicklung des physiologischen Fachschrifttums (Periodica) in den letzten 150 Jahren. Centaurus $\underline{4}$, 63-66 (1955)

SEN, B.R.: The Basic Freedom - Freedom from Hunger- FAO 1963

SCHEER, G.: Schriftenreihe Naturschutzstelle Darmstadt $\underline{4}$, 93-132 (1958)

SIOLI, H. (Ed.): Ökologie und Lebensschutz in internationaler Sicht. Freiburg i. Br. 1973

Statistisches Jahrbuch für die Bundesrepublik Deutschland. Hrsg. v. Statistischen Bundesamt, Wiesbaden, Stuttgart und Mainz 1966

Statistisches Jahrbuch der Deutschen Demokratischen Republik. Hrsg. v. d. Staatlichen Zentralverwaltung für Statistik. 11. Jahrg. Staatsverlag der DDR 1966

Yearbook of nordic Statistics - Nordisk Statistik Arsbok 1966. Published by the Nordic Council. Utgiven av Nordiska Radet. Stockholm 1967

ZISWILER, V.: Bedrohte und ausgerottete Tiere. Berlin-Heidelberg-New York 1965

VII. The Essence of Health and Its Maintenance

BLASIUS, W.: Erkenntnistheoretische und methodologische Grundlagen der Physiologie. In: LANDOIS-ROSEMANN: Lehrbuch der Physiologie des Menschen. Vol. II, 28th ed., p. 990-1011. München-Berlin 1962

BLASIUS, W.: Zur Geschichte der Reflexlehre unter besonderer Würdigung des Beitrages von Paul Hoffmann. Dtsch. Zschr. Nervenheilk. $\underline{186}$, 475-495 (1965)

BLASIUS, W.: Ganzheiten und Systeme und ihre Grenzen. Vitalstoffe $\underline{10}$, 152-155 (1965)

BLASIUS, W.: Raum und Zeit als Gestalten und Begriffe - Zu den mythischen Ursprüngen von Raum und Zeit in der Theogonie des Hesiod. In: Hestia 1965/66, p. 9-26. Bonn 1966

BLASIUS, W.: Rhythmus und Polarität - physiologische Analyse und erscheinungswissenschaftliche Deutung. Dtsch. Med. Wschr. $\underline{93}$, 1489-1495 (1968)

BLASIUS, W.: Die physiologischen Grundlagen der Übungsbehandlung. Zschr. Physik. Med. $\underline{1}$, 394-412 (1970)

Encyclopaedia Britannica $\underline{11}$, 295 (1963)

Der Gesundheits-Brockhaus. 2nd ed. Wiesbaden 1964

Der Große Brockhaus. 16th ed. Vol. IV. Wiesbaden 1954

GOETHE, J.W.v.: Die Natur - Ein Fragment aus dem "Tierfurter Journal 1782", Goethes Werke, Hamburger Ausgabe. Vol. XIII, 3rd ed., p. 45 ff. Hamburg 1960

HÖNEL, H.: Der Ausdruck des Willenstalentes. In: Festschrift zum 75. Geburtstag v. L. Klages. Ed. v. H. Hönel. Linz 1947

KLAGES, L.: Vom kosmogonischen Eros. 2nd ed. München 1926

KLAGES, L.: Brief über Ethik. In: Mensch und Erde - Sieben Abhandlungen. 5th ed. Jena 1937

KLAGES, L.: Vom Wesen des Rhythmus. 2nd ed. Zürich 1944

KLAGES, L.: Vom Wesen des Bewußtseins. 4th ed. München 1955

KLAGES, L.: Ausdrucksbewegung und Gestaltungskraft. In: Sämtliche Werke, Vol. VI: Ausdruckskunde. Bonn 1964

KLUGE, F., GÖTZE, A.: Etymologisches Wörterbuch der deutschen Sprache. Berlin 1948

Knaurs Gesundheitslexikon. München-Zürich 1951

MOMMSEN, H.: Die Gefährdung des Kindes in der modernen Zivilisation. In: Gefährdete Schöpfung. Homburg and Zürich 1970

IX. Authors Index

Alkmaion 2,35,69,90,170
Allan, E.C. 144
Aristotle 2,5
Aschoff, J. 50,51,52,53,54,55,56, 57,60,62

Baade, F. 167
Bach, J.S. 182
Bachofen, J.J. 171
Beethoven, L. van 182
Berger, H. 104,106
Bernhard, Cl. 3
Bertalanffy, L. v. 24,34,40,45, 162,166
Bertaux, P. 153
Blasius, W. 3,16,46,57,58,59,60, 66,70,72,73,74,76,78,82,96,101, 110,127,132,135,136,138,143, 148,149,151,153,156,157,160, 161,162,163,165,166,176
Boerhave, H. 2
Börnstein, R.L. 150
Boltzmann, L. 39
Broca, P. 102
Bruns, H. 155

Carus, C.G. 2,5,7,9,108,171,182
Clara, M. 99,102,103,104

Dandy, W.E. 103,104
Darwin, Ch. 32
Descartes, R. 5,16
Dürer, A. 147,148

Eichendorff 119
Einhorn, A. 7
Elsasser, W.M. 14

Escherich, K. 178
Eysenck, H.J. 144

Fenn, W.O. 161
Frauchiger, E. 8

Galen 2
Galileo, G. 24
Gauer, O.H. 84,85,87
Glees, P. 102,103
Göpfert, H. 79
Goethe, J.W. v. 2,5,6,9,10,32, 123,144,171,182
Grützmacher, M. 98,99
Guilford, J.P. 144

Hagen, G. 27
Haller, A. v. 2
Heiss, E. 144
Helmholtz, H. v. 3,5,36
Heraclitus 2,5,9,19,49,70,169, 171
Hill, A.V. 40
Hippocrates 2
Hönel, H. 175
Hofstätter, P.R. 144

Johannson, J.E. 81,82,83

Kant 5
Katsuki, I. 97
Klages, L. 5,7,8,9,11,18,19,64, 65,66,67,69,92,93,94,108,109, 110,112,113,114,115,116,117, 120,121,122,123,140,141,142, 146,154,171,172,176,181,182, 183

Koch, Eb. 124,125,127
Kopsch, Fr. 99,102
Kretschmer, E. 25,129,143
Kugler, J. 105

Lamparter, H. 129,130,132,143
Landolt, H. 150
Liebig, J. v. 46
Ludwig, C. 3,5
Lübbert, H. 144

Mayer, Robert 3,5,26,36,46
Mommsen, H. 178,179
Mozart, W.A. 182
Müller, Joh. 13,20

Neumann, J. v. 14
Newton, Fr. 5
Nietzsche, Fr. 1,9,10,11,49,171,182
Novalis 36,147,171

Oken, L. 2,5
Ortega y Gasset, J.
Ostwald, W. 7,128

Palágyi, M. 7,140
Panse, Fr. 107
Paracelsus 2
Plato 5,118
Poiseuille, L. 27
Prigogine, J. 40,44
Protagoras 5,171

Ranke, O.F. 84,87,88
Rashevsky, N. 40
Rauber, A. 99,102
Rein, H. 99,102
Repges, R. 15,60,124
Rohracher, H. 106
Rollmann, W. 125
Rothschuh, K.E. 3,152

Schäfer, A. 152
Scheer, G. 153,154,155,157
Schelling, F.W. 2,5
Schneider, M. 99,102
Scholl, R. 129,130
Schrödinger, E. 44
Sen, B.R. 157
Soili, H.

Tobler, G.Ch. 183

Ulmer, F. 58

Vasari, G. 148
Vollmer, O. 129

Walter, W.G. 106,107
Weizsäcker, V. v. 142
Wellek, A. 143
Wever, R. 54,56,57
Wiener, N. 14
Wieser, W. 14

Ziegenhain, L. 124,128
Ziswiler, V. 155

X. Subject Index

abstractive thinking 20
activity 50
- quantity 58
adaptation 71,80
-, muscular 81
amino acids 15
analogies 13
analytical-reductionist interpretation of vital processes 8
anatomy 93
animation 35
apnea 44
archetypal images 4
archetypes 6
art 123
articifical light-darkness cycle 50
association areas 102
- center 93
autonomic nervous system 88
autoregulation 43

beauty 181
bioelectric potentials 43
biological clock 54
- equilibrium 39
biology 14,94,152
body 110,175
bones 69
Broca motor speech center 103

carbon compounds 30
cardiac activity 58
cardiovascular 179
cartilage 69
causal theory 94
cellular organization 29
chain of causes 18
chaos 172

chemodynamic system 37
circulation 86
closed systems 41
cognition 93
cognitive act 141
coincide and its meaning 6
color discrimination 128,134,138
- - ability 136
- selection 144
- - tests 139
color-form-preferrers 135
color-preferrers 132,135,139,146
compensation 44
conditions of life 32
connection between body and soul 7
conscious perception 140
constitutional types 25,143
creativity 146
- tests 130
cyclothymics 143

difference between organismic and extra-organismic nature 28
disease 169
diurnally and nocturnally 55
dynamic equilibria 40
- equilibrium 41,42,43,44,62

EEG 106,107
ego 11
emotion 175,176
energy cycle 46
- transformations 37
- uptake and release 62
entropy 32,35,39,41,44,45,46,62
equilibrium systems 39
Eros 172

excitability 34
exercise 71,79
- therapy 89
experimental psychology 94
exponential growth

fever 180
form and color perception 140
form-color-preferrers 135,139
form-preferrers 132,135,139,146
forms 10

Gaia 172
Gestaltlehre 5
Gestalt theory 4
glands 69
growth curve in a closed population 163

health 169-185
hearing 95-122
heart 57,69
- muscle 83
heat economy 179
heredity 15
holism 5
holistic 2,3,20,70,91
- image of the universe 10
- world view 4
hygiene 152,169

images 10
information theory 14
intensity of illumination 53
internal clock 62
interplay of muscles and nervous system 71
interpretation of vital processes 15
ixothymics 143

joints 69

language 92-122
- centers 103
law of conservation of energy 36
- of constant total heat 36
life force 12
ligaments 69

limits of a natural scientific treatment of living processes 18
linguistic concept 113
- meanings 112,113
linguistics 112,170
localization of linguistic understanding 93
lungs 69

macromolecules 14,15
mathematical methods 26
meaning and conception words 116
- of life 10
medicine 152,169
memory 93,100
metabolism 179
methods of the natural scientist 11
mobility 175
- of the spirit 175
morphology 21
-, Goethe's 6
motor and sensory cortical fields 102
movedness 175
- of the soul 182
movement 16,17,69-91,175,176, 177,180
muscle spindles 75,76,91
muscles 69

natural laws 23
- scientific reasoning 18
nervous system 69,179
neurology 93,94

ontogeny 34
open systems 30,40,41,43,44
organismic and extra-organismic world 30
oxygen debt 44

periodic processes 50
periodicity 62
phenomenology 4,5,36,94
- (Wesenslehre) 12
phenomenological view of language 107
philosophy, natural 3,9,13
phylogenetic evolution 32

physical activity 180
physiology 1,2,21,93,94,169
polar character of life 9
polarities 4,6,9,13,35
polarity 18,49,66,123,140,145, 171,172,173,174,179,181
polarity, soul-body 108,113,143
population curves 166
- growth 167
practice 80
principles 70
problems of form 24
protoplasm 29
psychic factor 12
psychology 17,18

quantitative thinking 147

rationalism 16
reality of images 18
reductionist 2,3,14,15,20
- system 4
reflex arc 16
- theory 15
renaissance 111
respiration 86
rest 50
rhythm 49,67,171
rhythm and polarity 64
rhythmic movement 171
robot 14
romanticism 171

schizothymics 143
schizothymic or cyclothymic attributes 130,131
science, causal-analytical 5
-, natural 3,8,9,36
self-diagnosis 130
self-regulation of metabolism 40
sleep-waking periodicity 52
soul 18,110,114,171,172,173,175, 181
- and spirit of language 107

speech 95-122
- center 93
spirit 110,114,141,171,172
steady state 30,40
stereoscopic vision 131,134,138
stream of life 33

technology 152
temperature fluctuations 54
theory of life based on natural science 4
- -, holistic 5
- -, natural philosophical 4,5
- -, natural scientific
- -, reductionist 10
- of forms 5
thermodynamic system 37
thermodynamics 36
thinking, biocentric 5
-, logocentric 5
- in numbers 147
traffic deaths 162

uniqueness of man 48
universal life in the cosmos 48
universality of life in the cosmos 10

vascular system 69
viscera 69
vision 123-146
vitalism 5
vitalists 12

waking and resting phases 56
war dead 160
way of life 178
Wesenslehre 5
world population 153,160

Zeitgeber 56,57,61,63,67
-, innerer 62

The Study of Time
Proceedings of the First Conference of the International Society for the Study of Time, Oberwolfach (Black Forest), West Germany. Editors: J. T. Fraser, F. C. Haber, G. H. Müller. 65 figures. VIII, 550 pages. 1972. Cloth DM 77,–; US $ 31.60. ISBN 3-540-05824-9

As this volume is organized according to various themes, it comprises a multidisciplinary survey of the most significant modern interpretations of the concept of time. The unusual proceedings it reports yield a variety of insights which may be expected to leave their mark on theory and practice relating to time.

The Study of Time 2
Proceedings of the Second Conference of the International Society for the Study of Time, Lake Yamanaka, Japan. Editors: J. T. Fraser, N. Lawrence. 80 figures, 4 tables. VII, 486 pages. 1975. Cloth DM 57,–; US $ 23.40. ISBN 3-540-07321-3

This volume contains the proceedings of the Second World Conference on the Study of Time, which took place in the summer of 1973. The range of the topics covered is very wide, extending from biological and physical questions, to philosophical topics, political philosophy, and sociological questions. Contributions of a special session, Timekeepers and Time, conclude the volume; the articles in this section are devoted to historical topics, and include many attractive illustrations of old or unusual clocks.

Scientists in Search of Their Conscience
Proceedings of a Symposium on "The Impact of Science on Society" organized by the European Committee of The Weizmann Institute of Science, Brussels, June 28–29, 1971. Editors: A. R. Michaelis, H. Harvey. With contributions by numerous experts. 18 figures. XIII, 230 pages. 1973. DM 46,–; US $ 18.90. ISBN 3-540-06026-X

This symposium, organized by the European Committee of the Weizmann Institute of Israel, debates the responsibility of scientists for the uses to which society puts their findings.

Prices are subject to change without notice

Springer-Verlag Berlin Heidelberg New York

J. C. Eccles: Facing Reality
Philosophical Adventures by a Brain Scientist · 36 figures. XI, 210 pages.
1970 (Heidelberg Science Library, Vol. 13) DM 25,–; US $ 10.30.
ISBN 3-540-90014-4

The title of this book refers to personal reality and the book is concerned with the attempt of each person to face up to his own personal existence as a unique conscious self, recognizing at the same time how dependent this is upon the functioning of his brain and on his evolutionary origin.

To Live and To Die: When, Why, and How
Editor: R. H. Williams · 20 figures. XIX, 346 pages. 1973. Cloth DM 35,–;
US $ 14.40. ISBN 3-540-06220-3

In this book distinguished contributors address themselves to the problems of living and dying in our advanced technological society. Developments in science and medicine, population density, new life-styles and the erosion of criminal justice have called traditional standards of morality into question and led many to seek new value systems for a rapidly changing society. A number of concepts discussed in this book run counter to religious tenets and existing laws and are the subject of heated debate among laymen, as well as within professional communities.

Uprooting and After... Editors: C. Zwingmann, M. Pfister-Ammende
16 figures, 42 tables. XI, 361 pages. 1973. Cloth DM 68,–; US $ 27.90.
ISBN 3-540-05516-9

The present volume summarizes research in the areas of uprooting and resettlement. This is followed by carefully chosen examples of socio-psychological field work with resettlement groups in the various nations. This form of presentation provides an inner continuity in a field of study and research which by its nature is multidisciplinary and transcultural.

E. Straus, M. Natanson, H. Ey: **Psychiatry and Philosophy**
Editor: M. Natanson. Contributions are taken from Psychiatrie der Gegenwart 1/2. Translators: E. Eng, S. C. Kennedy · XII, 161 pages. 1969.
Cloth DM 46,–; US $ 18.90. ISBN 3-540-04726-3

Prices are subject to change without notice

Springer-Verlag Berlin Heidelberg New York